Through Silicon Vias

Materials, Models, Design, and Performance

T0315213

Through Silicon Vias

Materials, Models, Design, and Performance

Brajesh Kumar Kaushik • Vobulapuram Ramesh Kumar
Manoj Kumar Majumder • Arsalan Alam

CRC Press
Taylor & Francis Group
Boca Raton London New York

CRC Press is an imprint of the
Taylor & Francis Group, an **informa** business

CRC Press
Taylor & Francis Group
6000 Broken Sound Parkway NW, Suite 300
Boca Raton, FL 33487-2742

First issued in paperback 2020

© 2017 by Taylor & Francis Group, LLC
CRC Press is an imprint of Taylor & Francis Group, an Informa business

No claim to original U.S. Government works

ISBN 13: 978-0-367-57454-3 (pbk)
ISBN 13: 978-1-4987-4552-9 (hbk)

Visit the Taylor & Francis Web site at
http://www.taylorandfrancis.com

and the CRC Press Web site at
http://www.crcpress.com

Contents

Preface

The conventional two-dimensional integrated circuit (2D IC) packaging technique has almost reached its maximum profitable limit and is no longer useful for future IC integration. With the advancement in technology, the density of dies keeps increasing, and, therefore, the number of input/output (I/O) pins increases exponentially according to Rent's rule, and the interconnect length also increases to communicate between the dies. Therefore, due to the limited number of I/O pins and longer interconnects between dies, 2D IC integration offers lower bandwidth and thus degrades system performance. Recent advances in semiconductor technology offer vertical interconnect access (via) that extends through silicon, known as through silicon via (TSV). Compared to the conventional wire bond, TSVs offer higher bandwidth and density with low latency, and power dissipation, thereby enabling higher integration density and superior system performance. The use of TSVs is the only way to overcome the difficulties of 2D packaging issues while extending the momentum of Moore's law for future very-large-scale integration (VLSI) technology by using the advanced packaging chips named as three-dimensional (3D) chips.

Development of a reliable 3D integrated system is largely dependent on the choice of filler materials used in TSVs. Although several researchers demonstrated that copper (Cu) is a suitable filler material, but recently graphene-based nano-interconnects have rapidly gained interest to replace Cu. Graphene-based nano-interconnects can exhibit unique electrical, thermal, mechanical, and chemical properties. The sp^2 bonding in graphene is stronger than the sp^3 bonding in diamonds. The higher current-carrying capability, long ballistic transport length, higher thermal conductivity, and mechanical strength are responsible for their exciting prospects in the area of advanced packaging techniques.

This book provides a comprehensive review of theory behind TSVs, covering the most recent advancements in materials, models, and design. Furthermore, depending on the geometry and physical configurations, different electrical equivalent models for Cu-, carbon nanotube (CNT)-, and graphene nanoribbon (GNR)-based TSVs are presented. Based on the electrical equivalent models, the performance comparison among the Cu-, CNT-, and GNR-based TSVs is also discussed. The organization of the book is as follows: Chapter 1 introduces the current research scenario in 3D technology and packaging techniques. Chapter 2 discusses the structure, properties, fabrication techniques, and different filler materials of TSVs. Chapter 3 presents the scalable electrical equivalent model of Cu-based TSVs. Additionally, a novel approach is discussed to extract the parasitic parameters of Cu-based TSVs. Chapter 4 provides a brief review of CNTs and the performance comparison of Cu- and CNT-based TSVs. Chapters 5 and 6 discuss the mixed CNT bundled TSVs and GNR-based TSVs, respectively, along with the performance comparison of Cu- and CNT-based TSVs. Chapter 7 is dedicated to TSV liner materials and their impact on performance. Finally, Chapter 8 introduces the modeling of TSVs using the finite-difference time-domain technique.

Brajesh Kumar Kaushik
Vobulapuram Ramesh Kumar
Manoj Kumar Majumder
Arsalan Alam

Authors

Brajesh Kumar Kaushik received his BE degree in Electronics and Communication Engineering from the Deenbandhu Chhotu Ram University of Science and Technology (formerly *Chhotu Ram State College of Engineering*), Murthal, Haryana, in 1994; his M.Tech degree in Engineering Systems from Dayalbagh Educational Institute, Agra, Uttar Pradesh, in 1997; and his PhD degree under the All India Council for Technical Education-Quality Improvement Programme (AICTE-QIP) scheme from the Indian Institute of Technology Roorkee, Roorkee, Uttarakhand, India, in 2007. He served at Vinytics Peripherals Pvt. Ltd., Delhi, from 1997 to 1998 as the Research and Development Engineer for microprocessor-, microcontroller-, and DSP processor-based systems.

Dr. Kaushik joined the Department of Electronics and Communication Engineering, Govind Ballabh Pant Engineering College, Pauri Garhwal, Uttarakhand, India, as a Lecturer in July 1998, where he also served as an Assistant Professor from May 2005 to May 2006 and as an Associate Professor from May 2006 to December 2009. He is currently serving as an Associate Professor in the Department of Electronics and Communication Engineering, Indian Institute of Technology Roorkee. His research interests include high-speed interconnects, low-power very-large-scale integration (VLSI) design, CNT-based designs, organic thin-film transistor design and modeling, and spintronics-based devices and circuits. He has published extensively in several national and international journals and conferences of repute. Dr. Kaushik is a reviewer of many international journals belonging to various publication houses such the Institute of Electrical and Electronics Engineers (IEEE), the Institution of Engineering and Technology (IET), Elsevier, Springer, Emerald, and Taylor & Francis. He has delivered many keynote addresses in reputed international and national conferences. He holds the position of Editor and Editor-in-Chief of various journals in the field of VLSI and Microelectronics. He is a senior member of IEEE and has received many awards and recognitions from the International Biographical Center, Cambridge, England. His name has been listed in *Marquis Who's Who in Science and Engineering*® (10th Anniversary, 2008–2009 Edition) and *Marquis Who's Who in the World*® (26th Edition, 2009).

Vobulapuram Ramesh Kumar received his B.Tech in Electronics and Communication Engineering from Acharya Nagarjuna University, Guntur, Andhra Pradesh, in 2007, M.Tech degree in VLSI design from the National Institute of Technology, Hamirpur, Himachal Pradesh, India, in 2010, and PhD degree from the Indian Institute of Technology Roorkee, Roorkee, Uttarakhand, India, in 2015. He has authored/coauthored over 20 publications in reputed journals and conferences. He has coauthored a book titled *Crosstalk in Modern On-Chip Interconnects-A FDTD Approach* (Singapore, Springer, 2016).

Dr. Kumar worked as an Assistant Professor in the Department of Electronics and Communication Engineering at Gandhi Institute of Technology and Management (GITAM) University, Hyderabad Campus, Andhra Pradesh, from June 2010 to December 2011. He is currently working as an Associate Professor in the Department of Electronics and Communication Engineering at KL University (officially the Koneru Lakshmaiah Education Foundation), Guntur, Andhra Pradesh, India. His current research interests include time-domain numerical methods to approach fast transients characterization techniques, modeling of VLSI on-chip interconnects, graphene-based nano-interconnects, and through silicon vias.

Manoj Kumar Majumder received his B.Tech and M.Tech degrees from West Bengal University of Technology, Bidhan Nagar, West Bengal, India, and Indian Institute of Engineering Science and Technology (IIEST), Shibpur, West Bengal, India, in 2007 and 2009, respectively. He was awarded his PhD degree from the Department of Electronics and Communication Engineering, Indian Institute of Technology Roorkee, Roorkee, Uttarakhand, India, in 2015. He was a Lecturer in the Durgapur Institute of Advanced Technology and Management, Durgapur, West Bengal, India, from July 2009 to July 2010. He is currently associated with the academic activities in Electronics and Communication Engineering Department, Techno India University, Kolkata, West Bengal, India.

He has authored over 20 papers in peer-reviewed international journals and over 30 papers in international conferences. He has coauthored a book titled *Carbon Nanotube Based VLSI Interconnects: Analysis and Design* (New York: Springer, 2014) and also a book chapter entitled "Fabrication and modelling of copper and carbon nanotube based through-silicon via" in *Design of 3D Integrated Circuits and Systems* (Boca Raton, FL: CRC Press,

2014). His current research interests include the area of carbon nanotube and graphene nanoribbon-based VLSI interconnects and vias. His name has been listed in *Marquis Who's Who in the World* (33rd Edition, 2016).

Arsalan Alam received his B.Tech degree from Zakir Hussain College of Engineering and Technology, Aligarh Muslim University, Aligarh, Uttar Pradesh, India, in 2011 and his M.Tech degree in Microelectronics and VLSI from Indian Institute of Technology Roorkee, Roorkee, Uttarakhand, India, 2015. He is currently working as a Research Associate in King Abdullah University of Science & Technology, Thuwal, Saudi Arabia. His research areas are time-domain numerical methods for analyzing the performance of VLSI on-chip interconnects and modeling of through silicon vias.

1

Three-Dimensional Technology and Packaging Techniques

1.1 Introduction

A new set of solutions is clearly required for the packaging technology to keep up the demand for high integration and low power consumption. Chip stacking has arguably emerged as the technology that caters to the need of both the advanced packaging techniques and the technology scaling. Numerous innovative designs of chip stacking have been proposed, and some of them have successfully been implemented for commercial purposes for many years. However, in the past, other techniques have been implemented to connect the stack chips. Some of these techniques are wire bonding, ball grid arrays (BGAs) or flip chip bumps, wafer-level packaging (WLP) of chips, and so on. All these techniques were able to provide improved functionality in a single package and were successful in replacing the earlier printed circuit board (PCB) connections. Additionally, stacking packaging techniques provide better electrical connectivity leading to reduce power consumptions and improve reliability of the system [1].

In recent years, a new packaging technique for chip stacking called the three-dimensional integrated circuit (3D IC) has been introduced to keep up the demands of power reduction and high integration. In 3D ICs, the dies are vertically stacked [2] using through silicon vias (TSVs) [3]. Before discussing the latest trends in packaging, it is important to have some knowledge about the conventional packaging techniques.

1.1.1 Conventional Packaging Techniques

In the early 1950s, single chip packages were introduced based on the hermetic can package technique. In this technique, a small circular hermetically sealed metallic can was formed with 2–10 pins protruded outwards [4]. An example of hermetic can package is shown in Figure 1.1. The pin count was not an issue, and they were used to package the single transistor chips. In the early 1960s, integrated circuits (ICs) were introduced and henceforth started the demand for higher pin count. The first monolithic IC was developed in 1961

FIGURE 1.1
A hermetic can package.

and it required six pins. The master–slave flip-flop was developed in 1964 and it had 11 pins. The J–K flip-flop introduced in 1966 demanded 16 pins. A typical arithmetic logic unit (ALU) in 1967 required 36 pins. As a result, the metal cans could no longer keep up with burgeon demand for pin numbers, and their future was soon obliterated.

The dual-in-line package (DIP) that allowed through holes in the circuit boards was developed to cater for the need of increasing pin counts. This technique has the term "dual" because it has two rows of leads on two sides. Figures 1.2 and 1.3 show the examples of DIPs with 14 and 40 pins, respectively. Soon, the DIP could no longer provide for the aggressive increase in demand for the IC pin count, and a new technique named pin grid array (PGA) was introduced. The PGA is also a through hole-based package technique, in which pins are arranged in an area array pattern. This technology can easily accommodate more than 100 pins.

In the mid-1970s, there was a transition towards surface mounting of the package that leads to the development of solder pads. The solder pads replaced the prior technique of slotting the pins into holes. This transition opened the doors for more densely packed components on boards. Packaging technology soon had leads being present at the periphery. DIPs are regarded as the early surface mount packages, in which the butts are attached to the board and the leads cut short. In the 1970s, a new generation of packages having leads on all four sides and mounted to the board surface was introduced. Such a package technology was given the term "flat pack." In the 1980s, a new package technology having attached points around the periphery, with or without leads, and typically square in shape was developed. It was recognized by the term "chip carrier."

In recent years, the interconnection density, technology integration, cost, integration density, form factor, and latency have started to play an important role. Interconnection density is the measure of the number of vertical connections

FIGURE 1.2
A DIP with 14 pins.

FIGURE 1.3
A DIP with 40 pins.

per unit area. Technology integration is the capability of integrating different process technology-based chips or dies. Integration density is the capability of a technique to incorporate multiple functionalities within a volume. Form factor is the measure of the final size of the product per functionality, and the latency of vertical interconnects is related to performance. New techniques were introduced which were focused on excelling in one or more of these parameters.

Several heterogeneous chips in a package can be accommodated adjacent to each other through a technique called multichip modules (MCMs). This technique is low in cost; however, due to large interconnect lengths and area, it scores rather unsatisfactorily in other parameters. The chip stack technique, in which the chips are interconnected via wire bonds, does not excel in any parameter but provide satisfactory results for all parameters. In comparison to MCMs, they are better in terms of form factor, integration density, and performance, because the wire bonds provide better electrical characteristics. However, scalability issues are associated with wire bonds. The peripheral input/output (I/O) chips can only be linked via wire bonds, and therefore, interconnection density is limited for wire bond chip stacking technique. An alternative to wire bonds is the solder ball. The technique in which various chips in a stack are interconnected via arrays of solder balls is called the BGA stack. The solder balls provide better electrical characteristics than the wire bonds. Although they are not as cheap as the wire bonds, they provide better technology integration. WLP is a packaging technology in which vias through the substrate are used to interconnect the various dies. The WLP provides the better form factor and integration density compared to the BGA.

1.1.2 Limitations

The demand for light weight, low power, higher performance, greater functionality, higher integration density and compact packaging technology
is drastically increasing with the passage of time. The conventional packaging techniques are gradually becoming incompetent to provide for the
ever-increasing demands of the electronic industry. Keeping pace with the
Moore's law, the chip density doubles in every 2 years, and the bandwidth
requirement is doubling every 3 years; however, the existing package
techniques are not able to keep pace with the demand. Therefore, novel packaging technologies are required that can provide the high chip density and
bandwidth to keep up the pace with Moore's law. The conventional techniques
are inept in providing efficient solutions for heterogeneous integration of
logic, radiofrequency, sensor and bio-MEMS ICs. A two-dimensional IC (2D IC)
process for system on chip (SoC) results in bigger chip sizes, additional process steps for dissimilar functional blocks, and longer interconnects. The
length of the interconnect increases due to an increase in chip functionality
and chip scaling. Accordingly, the interconnect capacitance and resistance
also increase, resulting in higher latency, increased power consumption and
crosstalk. The conventional packaging techniques seem to be slowly disap-
pearing from manufacturing industries and are being replaced by a new set
of packaging techniques.

1.1.3 Recent Advances in Packaging Technology

The best way to move toward the "More-than-Moore" technologies is the
3D IC packaging, in which the dies are vertically stacked [2]. Overall, the 3D
integration can be primarily divided into two approaches:

1. *Monolithic approach:* It involves sequential processing of device. The
   front-end processing, through which layers of device are built, is
   repeated numerous times on a single wafer to produce multiple
   active device layers prior to the back-end processing, which routes
   the devices together through the interconnects.
2. *Stacking approach:* It includes three different stacking methods—
   die-to-die, die-to-wafer, and wafer-to-wafer. In this approach, each
   layer is processed separately with the help of conventional techniques
   of fabrication. These processed layers are assembled together with
   the help of bonding technology to prepare a 3D IC.

Due to the use of conventional fabrication techniques, the stacking approach is
more practical than the monolithic approach. However, with the advancement
in fabrication technology, the monolithic approach is expected to gain more
importance. A number of 3D stacking technologies, such as microbumps, TSVs,
wire bonded, and contactless (inductive or capacitive), are available. Among all

these technologies, the TSV-based 3D IC offers the highest integration density and is considered the most promising vertical interconnect technology.

A TSV-based 3D IC is a single package containing vertical stack of dies and allows the die to be vertically interconnected with another die [3]. TSVs can provide a shorter signal path that exhibits superior electrical characteristics in terms of reduced resistive, inductive, and capacitive components [5]. Moreover, the TSVs in 3D ICs carry the advantage of having fine pitches. This technology offers the highest performance, integration density due to smallest volume, interconnection density, and smallest form factor. Because it is a relatively newer technology, it is quite expensive. However, continuous research and increasing demand will only make it cheaper and hence more suitable as a packaging technique for future ICs. Therefore, the primary focus of current research is to design reliable and cost-efficient TSVs.

In TSV formation, wafer thinning and die bonding are the primary process steps that are followed for 3D stacking. The impact of TSVs around its surrounding areas is reduced with the help of wafer thinning process. In the bonding of TSV, the scaling rate of TSV dimensions is not expected to be at par with feature size. This is because during bonding, the alignment tolerance limits the scaling of TSVs. However, the monolithic approach for 3D integration allows TSV scaling to be at par with feature size. This is because the connections are made from local wires.

Moreover, using the 3D technology, one can integrate different homogeneous materials (silicon, III–V semiconductors, carbon nanotube, graphene, etc.) and applications (memory, logic, mixed signal, radiofrequency, optoelectronics, etc.) on a single chip [6].

1.2 Packaging Techniques of Future ICs

The conventional packaging technique faces major challenges in the form of limited number of pins and lower bandwidth. Recently, a new IC integration technique called silicon interposer technology or 2.5D was reported [7]. The packaging of a 2.5D IC facilitates heterogeneous integration of logic, radiofrequency, memory, sensor, bio-MEMS ICs, in which the dies are fabricated at the optimum technology process. The dies are placed side by side on the top of a passive silicon interposer. This silicon interposer contains horizontal interconnects and TSVs to provide routing between the dies and the external power supplies, respectively. The 2.5D technology offers better thermal and testing management. Later on, another new integration technique called 3D IC was reported. The TSV-based 3D IC packaging technology offers the vertical integration of heterogeneous dies and routing between the stacked dies by means of TSVs. While holding the advantages of 2.5D ICs, the 3D ICs have additional benefits such as smaller footprint and form factor, which are essential for applications such as in wireless and portable devices. The 2.5D and 3D packaging techniques offer the most promising platform for integration of heterogeneous materials and dies on a single chip while moving a step ahead for implementation of More-than-Moore technologies.

The reduced feature size has been a solution to carry on the Moore's law since a long time. However, with time, the reduced feature size beyond a limit causes serious concerns on reliability, performance, and cost of the devices. Therefore, stacking of chips on top of each other, resulting in 3D integration, has provided a solution to increase the integration density of devices. However, stacking 3D integration in itself has issues and may not be as simple as it may sound. There is a fundamental difference between stacking of transistors and decreasing their size, to achieve higher device integration density. The power per transistor is the same for a given feature size [8]. Therefore, at the same technology node, more number of transistors per unit area through 3D integration also increases the power requirements. Moreover, the 3D stacking of layers also increases the chip area utilized, resulting in higher cost.

Various companies, including Samsung, have reached a consensus that 3D IC technology will soon be popular at the market level. It is important that TSV-based 3D ICs soon become a reality for market-level manufacturing. This is due to the reason that the 3D IC replaces conventional input/output (I/O) pins by TSVs to provide an exceptionally higher bandwidth. The 3D IC provides a platform to integrate different process technologies such as dynamic random access memory (DRAM) on complementary metal–oxide–semiconductor as well as enhance the on-chip storage capacity. Moreover, it reduces the power requirement and provides improved performance due to the reduced need of interconnects and I/O drivers linked to off-chip memories. There are many more significant benefits of TSV-based 3D ICs; however, they do face some challenges. The thermal management of multiple power hungry dies stacked together in a compact volume is a major issue. Placement of high-power demanding processor core layers near the heat link may be a solution to the heat issues. The TSV-based 3D IC also requires the close collaboration of designers and vendors to provide standards for different component interfaces, such as memory–processor interface standards. For short-term solutions, the automation of electronic design and design tools requirement would suffice. However, in long term, heavy investment is required for designing these tools and providing standardization. A good vision is also required to make the TSV-based 3D IC a success. This is because 3–5 years prior to manufacturing, product planning is initialized. It is important that the fabs of companies are equipped to handle a 3D process; however, it requires the risk of very high costs before and during manufacturing. A more serious question is how much would the users pay for additional benefits of TSV-based 3D integration. A general opinion regarding 3D integration is that although there are issues with TSV-based 3D integration, there are absolutely no fundamental issues [8,9]. Therefore, it is expected that TSV-based 3D IC will definitely be market ready soon. The latest packaging techniques are discussed in Sections 1.2.1 through 1.2.5.

1.2.1 Silicon Interposer Technology

The term "interposer" has got its origin from a Latin word *interpōnere*, meaning "to place in between." As the name suggests, an interposer is placed in between the packaging substrate and dies to provide electrical interface routing. The objective of an interposer is to electrically reroute a terminal to different dies or to extend a connection to broader areas via wider pitch [10]. The ever-increasing need for miniaturization, higher performance, and multifunctional microelectronic devices is motivating the microelectronic industry to move toward higher integration density of interconnects and data bandwidth provided with good signal integrity. To provide a solution to these increasing demands, silicon interposers have emerged. An interposer is a dual-sided die containing no active components and is used to route chips to substrate or chips to one another [11]. The interposers allow the placement of dies next to each other. They provide shorter interconnections between the chip and the substrate, thus improving the electrical performance of the system by reducing the power dissipation and propagation delay.

The structure of a silicon interposer is shown in Figure 1.4. It can be observed that the silicon interposer enables routing between the dies and connects the overall package to PCB. Unlike the 2D IC technology, the 2.5D IC technology has an interposer sandwiched between the stacked chips and the substrate. The interposer, with the help of TSVs, connects the surfaces at its top and bottom, and also interconnects the various adjacent chip stacks. Microbumps and C4 bumps are used to connect the interposer to the chips and the substrate, respectively. The packaging of the 2.5D IC facilitates heterogeneous integration. The dies are placed side by side on the top of a passive silicon interposer. This silicon interposer contains horizontal interconnects at multiple layers to provide routing between the dies. Additionally, it contains TSVs to provide the connection between the dies and the packaging substrate. The 2.5D technology demonstrates better suitability for high-performance applications as this technology offers better thermal and testing management. However, the addition of interposers to ICs comes with a drawback of additional design complexities and fabrication issues for designers and manufacturers.

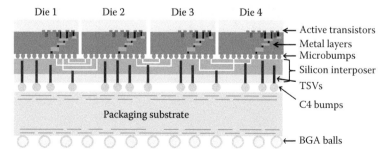

FIGURE 1.4
Structure of 2.5D IC. Stack dies are placed on top of a silicon interposer.

A 3D IC without interposer faces several challenges, such as thermal management and integration of different process technology chip. TSV-incorporated interposers allow the designers to take advantage of chip scale interlinked configurations, excluding the issues associated with 3D ICs without interposers. The usage of the interposer enables the circuitry of chips to be within the device package, thereby diminishing the requirements of guard rings and electrostatic discharge (ESD) protection circuits [12]. Interposers also help in easy routing between chips, whereas other requires complex designing procedures to satisfy the interface compatibility issues.

1.2.2 Through Silicon Vias

TSV provides integrated 3D packaging using the vertical stacking of chips. It can be referred to as a vertical electrical connection via (vertical interconnect access) that passes completely through a silicon wafer or die. A TSV-based 3D IC offers various advantages by integrating a heterogeneous system into a single platform as shown in Figure 1.5 [2]. TSVs are high-density interconnects compared to the traditional wire bonds [5]. They allow more interconnect lines between the vertically stacked dies, and hence exhibit higher speed and bandwidth. IC integration with TSVs is an emerging technology that forms multifunction high-performance ICs by providing significant benefits with improved performance, packing density, power consumption, and heterogeneous technology integration capabilities. Compared to the silicon interposer technology, the TSV-based 3D ICs have additional benefits

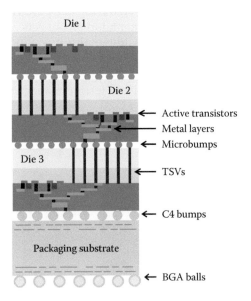

FIGURE 1.5
Placement of interconnects and TSVs in a 3D IC.

such as smaller footprint and form factor, which are essential for applications such as in wireless and portable devices [7]. Therefore, the primary focus of current research is to design reliable and cost-efficient TSVs. A typical TSV has a radius of 10 μm and a depth of 150 μm. One end of the TSV is connected to the on-chip metal layer, whereas the other end is connected to another die [7].

The TSV-based 3D technology provides the most promising platform to implement More-than-Moore technologies [7]. Using this 3D technology, one can integrate different homogeneous materials (silicon, III–V semiconductors, carbon nanotube, graphene, etc.) and applications (memory, logic, mixed signal, radiofrequency, optoelectronics, etc.) on a single chip.

Beyond TSV considerations, all the conventional design procedures and layout capabilities are not suitable for 3D integration. Using conventional designing, a few laboratories or companies had developed 3D-compatible design methodologies [13–15], but they were not commercially available. The well-known methodologies involved in realizing the 3D integration include face-to-face integration and back-to-back integration [5]. The ultimate success of TSV-based 3D ICs will rely on whether the huge potential of 3D integration can be translated into practical benefits by the manufacturing industry.

1.2.3 Hybrid Packaging Technique

A typical 3D IC offers the most promising platform for integration of heterogeneous materials and dies on a single chip while moving a step ahead for implementation of More-than-Moore technologies. However, due to the stacking of ICs on top of each other, thermal and stress management issues are very critical in 3D ICs. Therefore, the researchers are forced to find an alternative integration technique that provides lower footprint and higher bandwidth and offers better cooling options.

Keeping the advantages of both 2.5D and 3D ICs, and overcoming the critical thermal management issues in 3D ICs, a hybrid combination of 2.5D (silicon interposer) and 3D integration is utmost required. In the hybrid 3D IC, different dies are stacked together on top of each other and connected through TSVs, and the interposer routes the group of dies. All these components are mounted on the packaging substrate. Figure 1.6 shows a hybrid 3D IC in which the multiple dies are stacked on top of a silicon interposer. The TSVs shown in the IC substrate (Figure 1.6) are used to connect the vertically stacked dies, whereas the TSVs in the silicon interposer are used to connect external I/O pins to the integrated chips [7]. However, the hybrid 3D IC structure is more complex than TSV-based 3D or 2.5D structures. Addition of interposer to 3D IC results in increased area usage. Various process steps need improved solutions, such as verification and implementation of 3D and 2.5D ICs, planning of design tools, handling of thin wafers, testing, and thermal management. Sophisticated and state-of-the-art techniques are required to properly fabricate its structure.

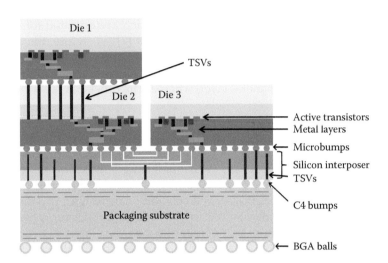

FIGURE 1.6
Structure of a hybrid 3D IC.

1.2.4 Silicon-Less Interconnect Technology

TSV-based silicon interposers have been adopted to provide enhanced performance and reduced power consumption. The interposer also reduces the stress on die having dielectrics of extra low-k. However, the TSV-enabled interposer increases the cost of developing the overall structure and creates a hindrance for market-based production. The TSV creation in interposers that involves barrier/seed metal deposition based on high aspect ratio, via filling, back grind, bond and debond of temporary wafer, chemical vapor deposition process for dielectric deposition, polish, and so on collectively make TSV-based interposers an expensive structure.

A glimpse of forward-looking development in 3D IC is the recently proposed silicon-less interconnect technology (SLIT). It has been recently applied in the development of FPGA integration [16]. In the conventional Xilinx-based FPGA module, the die is connected to the four layers of interconnect on the interposer. The Si-based interposer has C4 bumps and TSV for the purpose of routing the various signals, whereas in SLIT, the FPGA module is connected to the four layers of interconnect on Si without using TSVs. This is because of the careful removal of Si from selective areas and contact formation on the backside provided with important controlling of inline warpage [16]. The SLIT-based FPGA implementation is expected to provide improved electrical performance, and moreover, due to the removal of Si interposer, it will also lower the cost of the overall structure. The schematic of conventional Si interposer-based FPGA and SLIT-based FPGA are shown in Figure 1.7a and b, respectively.

Interconnects of 65 nm are produced on bulk Si as shown in Figure 1.7a. To assist during the backside etch, the bottom dielectric layer is preferred to have high selectivity. The top of the interconnect metallization layer is

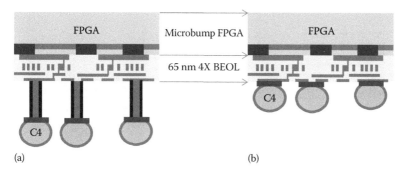

FIGURE 1.7
(a) Conventional Xilinx-based FPGA with Si interposer and (b) SLIT-based FPGA without interposer. (From Yang, Y. et al., *IEEE Photonics Journal*, 5, 2700611(1)–2700611(12), 2013.)

terminated in microbumps and 45 μm pitch pads. The interconnect wafer supports the stacked thinned diced FPGA. The microbump gap is undefiled and then over molded after reflow. The top of the die is then exposed by grounding down the molded compound. The Si present on the interconnect wafer is then removed till the M1 layer by etching [17]. Balls and pads are created for the purpose of the attachment. The removal of Si makes the wafer prone to warpage. To overcome this issue, a reinforcement layer is provided to balance out the stresses, and other stress-controlling measures are considered during processing. Because the TSVs are absent, the complex and expensive fabrication procedure for TSV development is not required. The handling of thin wafers, which is an important design requirement in conventional structures, is not necessary. Moreover, there is no need for additional backside reveal process in the SLIT technique.

1.2.5 Comparison of Different Packaging Techniques

With numerous packaging techniques developed over the years, it becomes important to study their various advantages and disadvantages and provide a comprehensive comparison. The comparison helps in selecting the best packaging technique based on individual requirements. It also paves the way for developing future packaging techniques by recognizing the shortcomings of available techniques.

A comparison of 2.5D, 3D IC, and hybrid 3D IC for various domains involving thermal reliability, device impact, design flow, testing, and cost is shown in Table 1.1. Thermal management is an issue in 3D IC, whereas 2.5D and hybrid 3D ICs have no such issues [18]. It is due to the fact that the heat generation in 3D IC is quite complex because 3D IC offers vertical stacking, whereas the 2.5D and hybrid 3D ICs offer planar packaging. Due to the higher integration density of 3D ICs, it becomes difficult to efficiently incorporate thermal management systems. Moreover, mechanical stresses are generated on regular heating and cooling due to IC components having

TABLE 1.1

Comparison of 2.5D, 3D, and Hybrid 3D ICs for Various Domains

Domain	2.5D IC	3D IC	Hybrid 3D IC
Thermal management and reliability	Good	Bad	Good
Device impact	None	Stress	None
Design flow	New design required/ challenging	Evolutionary	Evolutionary
Testing	Not challenging	New methods/ challenging	Not challenging
Cost	Lesser than 3D IC	High	Higher than 3D IC
Footprint	High	Low	High

different coefficients of thermal expansion. These stresses cause reliability issues, particularly in 3D IC, in which more dies of different materials are stacked together. Comparing the two ICs for device impact, it is seen that 3D IC presently should not be preferred. 3D IC has to overcome alarming challenges at the device level.

The TSVs are very large in dimensions compared to the transistors in 3D and hybrid 3D ICs. Due to their tremendous size, they easily impact the surrounding components or even the stacked die as a whole and disturb the uniformity in device characteristics. Therefore, careful evaluation of the position of their placement and challenging study of their impact are required before using them through the die. Such issues are absent in 2.5D IC because the TSVs are placed within the passive interposer. The 3D and hybrid 3D ICs also have design flow issues. Being relatively new technology, they may even require a new approach to the entire design flow. However, 2.5D IC has the benefit of requiring an evolutionary design flow. For testing considerations, 3D IC may require new test structures. Defect in die stacks is a very sensitive issue, in which even a single defect can be responsible for damaging the entire device. Each die may require a self-test to be assured of defect-free regions, but it would increase the complexity and cost. However, the 2.5D and hybrid 3D ICs do not demand any radical changes for testing. Moreover, in consideration of the cost of technology, the 2.5D IC is cheaper than the 3D and hybrid 3D ICs. The 3D IC offers the lowest footprint than the 2.5D and hybrid 3D ICs, which is essential for portable integrated chips.

3D IC can be realized in a variety of approaches. All these approaches can be broadly categorized in three different methodologies as shown in Table 1.2. Factors such as die size, fabrication/design constraints, materials, interconnection density, yield, and cost mainly decide the choice of appropriate method for 3D IC stacking.

The die-to-die methodology has the advantage of known good dies (KGDs) and considering multiple die sizes [19]. KGDs are those chips that are tested and then placed into packages. The die-to-die methodology has a disadvantage of low throughput due to the time-consuming die pick-and-place process.

TABLE 1.2

Comparison of Three Stacking Methodologies

Methodology	Schematic	Advantages	Disadvantages
Die-to-die		Multiple die sizes, KGDs	ESD, bonding and handling, low throughput
Die-to-wafer		Multiple die/wafer sizes, KGDs	ESD, accuracy of placement
Wafer-to-wafer		No ESD, high throughput, cheap	Design inflexibility, lack of KGDs

The ESD is the primary issue in this methodology. This is because the die pick-and-place process is susceptible to the potential difference between the device and the environment. The die bonding and handling in this approach is not very efficient. The die-to-wafer methodology also has the advantage of KGDs and choosing different die/wafer sizes. Moreover, it provides better throughput by using wafer-level processes. However, this approach is also susceptible to the ESD, and the process used for its die/wafer placement is not very accurate. The wafer-to-wafer methodology makes use of the monolithic wafer-level process, resulting in better reliability, high throughput and good yield, and hence lower cost. Modern manufacturing equipments eliminate the ESD issue from this approach. The drawback of this methodology is the absence of KGD, which may result in yield loss, and design inflexibility, especially for cases in which same sizes of different chips cannot be attained.

A comparative study between 3D system in package (SiP), functionality, 2D SoC, and 3D TSV for power consumption, performance, mass production cost, and ease of manufacturing is carried out and shown in Table 1.3. It can

TABLE 1.3

Comparison of 3D SiP, 2D SoC, and 3D TSV

Parameter	3D SiP	2D SoC	3D TSV
Functionality	+++	+	++
Power	+	++	+++
Performance	+	+	+++
Mass production cost	+++	++	+
Ease of manufacturing	+	++	+++

+++, ++, and + signify best, good, and worst, respectively.

be observed that the application of TSV in 3D TSV structures offers the best results for most parameters compared to SiP and SoC. However, 3D TSV is the most expensive technology.

1.3 3D Integrated Architectures

3D integration stacks multiple layers of dies with a high-density, low-latency layer-to-layer interconnect. It primarily provides four key benefits for the process architecture: (1) 3D placement and routing substantially that reduce the wire length that results in improved latency performance and reduced power consumption; (2) heterogeneous stacking as different types of components can be fabricated separately and the layers can be implemented with different technologies; (3) enhanced memory bandwidth, by placing memory on the cores of microprocessor with the help of multiple parallel TSV-based connections between the core and memory layers; and (4) small form factor that provides lower footprint advantage and higher packing density [20]. This section presents different 3D integrated architectures and their associated benefits.

1.3.1 3D Integrated Microprocessor

High-performance microprocessors contain billions of components per chip. It is quite difficult to integrate multiple components on a single die using global wiring. To circumvent this problem, a microprocessor can be designed using fine-granularity partitioning and stacking approach of 3D integration technology [20]. The designing approach is primarily dependent on the extremely small die-to-die interconnects with very high pitch density. Figure 1.8 shows the layout of 3D processor, in which each functional block is partitioned into multiple layers. A subarray-level partitioning methodology is used to divide bit lines and word lines across different layers. The corresponding local word line decoder of the original word line in 2D subarray is placed on one layer and is used to feed the word line drivers on different layers through the 3D vias. Using this methodology, the bit line length of the subarray as well as the number of pass transistors connected to a single bit line are reduced, which facilitates faster switching of the bit line array. It substantially reduces the length of global lines and the overall delay.

1.3.2 SRAM Array Integration

In 3D integration, the die-to-die vias are sufficiently dense to enable partitioning of the processor blocks. The array of static random access memory (SRAM) primarily includes a grid of SRAM cells, along with logic at the periphery such as sense amplifiers and row decoders. Loh et al. [21] reported

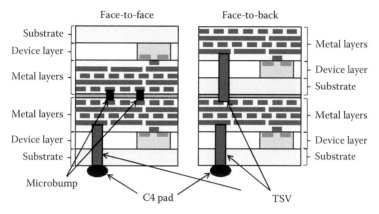

FIGURE 1.8
Layout of a 3D processor.

that in a 3D stack of two dies, the SRAM cells can be folded along the x- or y-axis, in a way that the bit lines or the word lines are split across the two layers or all odd bits are placed on one die and all even bits are placed on the other die. Figure 1.9 presents an SRAM array with vertically stacked 3D TSVs. Compared to the 2D integration, the length of either the word line or the bit line is approximately halved in the 3D integration depending on the partitioning. It results in reduced power consumption and array's footprint.

1.3.3 Network-on-Chip Architecture

Network on chip (NOC) is a general-purpose on-chip interconnection network architecture that can potentially replace the conventional design-specific global on-chip wiring and connect processor cores and different memory layers using TSVs (Figure 1.10). Typically, the processor cores communicate with each other using a packet-switched protocol [22]. Figure 1.11 shows a

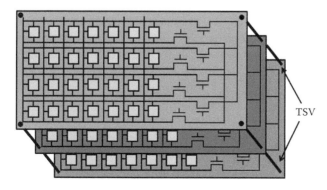

FIGURE 1.9
Integration of a 3D SRAM array (top view).

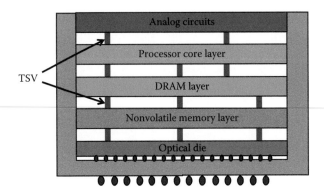

FIGURE 1.10
3D architecture with nonvolatile memory stacking and optical die stacking.

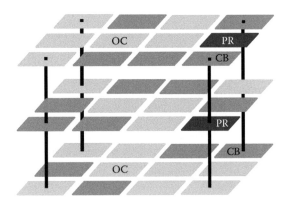

FIGURE 1.11
Processor core (PR), optical core (OC) and nearby cache banks (CBs) for 3D organizations.

processor core and nearby cache banks in 3D organization. In comparison to the 2D organization, more cache banks are reachable in 3D for a fixed number of hops. It results in significant reduction in the number of cache line migrations, which improves the average latency and reduces the overall power consumption and network connection compared to a 2D counterpart.

1.3.4 Wireless Sensor Network Architecture

The wireless sensors are composed of multiple components demonstrating a wide range of applications, including infrastructure monitoring and bio-medical implants. Recently, Lee et al. [23] fabricated a 1 mm³ general-purpose heterogeneous sensor node platform with a stacked multilayer structure to achieve smaller footprint. The structure contains five vertically stacked lay-ers, in which the wire bonding scheme is used for the electrical connectivity of the sensor system as shown in Figure 1.12.

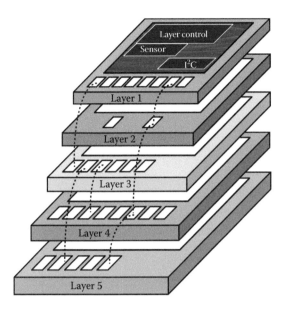

FIGURE 1.12
Schematic of the 1 mm³ sensing platform.

Due to pad count limitation, the ultralow power inter-IC is used for the interlayer connects that only require two wires of serial data for communication. In practice, the high-performance very-large-scale integration (VLSI) chip requires a higher number of I/O pins, but the wire bonding technique supports only a limited number of I/O pins. However, TSVs can sustain the demand of increased number of I/O pins. Therefore, TSVs are considered to be superior to the wire-bonded technique because they offer higher bandwidth and number of I/O pins.

1.4 Summary

The IC packaging industry has evolved from hermetic can package resembling vacuum tubes in the 1950s to potential TSV-based 3D ICs to meet the demand for More-than Moore technologies. A brief review of the growth of the packaging industry from the 1950s has been outlined in this chapter. The limitations of conventional packaging techniques to meet the demands of the end users and the packaging techniques of the future ICs have also been discussed. The advantages of TSV-based 3D IC to support integrated architectures such as integrated microprocessor, SRAM array integration, NOC design, and wireless sensor network architectures are also discussed. It has been demonstrated that TSV-based 3D ICs hold the potential to be the next-generation ICs. Its success depends on the efforts put together by the research world and manufacturing industries to transform its potential to real-life benefits for the end users.

Multiple Choice Questions

1. What is flat pack?
   a. A DIP technology
   b. A technology of solder pads
   c. A new generation of packages having leads on all four sides and mounted to the board surface
   d. None of the above
2. Which of the following techniques has the highest pin counts?
   a. DIP
   b. PGA
   c. Monolithic IC
   d. Solder pads
3. The MCM technique
   a. Integrates several heterogeneous chips in a package
   b. Requires low cost
   c. Requires more length and area
   d. All of the above
4. The advantage of the chip stack technique over MCM is
   a. High form factor
   b. High integration density
   c. High performance
   d. Both (b) & (c)
5. The advantage of TSV is
   a. Superior electrical connectivity
   b. Highest performance
   c. Highest integration density
   d. All of the above

6. The advantage of TSVs over conventional I/O is
   a. Higher bandwidth
   b. High on chip storage capacity
   c. Reduced power requirement
   d. All of the above
7. Which of the following techniques has the lowest footprints?
   a. 3D IC
   b. Hybrid 3D IC
   c. 2.5D IC
   d. Solder pads
8. Which of the following has the lowest power consumption?
   a. 2D SoC
   b. 3D SiP
   c. 3D TSV
   d. DIPs
9. The application of 3D TSVs is
   a. 3D integrated microprocessor
   b. SRAM array integration
   c. Wireless sensor network architecture
   d. All of the above
10. The methodology used for the 3D IC realization is
    a. Die to die methodology
    b. Die to wafer methodology
    c. Wafer to wafer methodology
    d. All of the above

Short Questions

1. Define the terms "interconnect density" and "technology integration."
2. Explain form factor.
3. What is chip carrier?
4. What is MCM?

5. What are the advantages and disadvantages of the MCM?
6. Explain about the BGA stack technique.
7. What is WLP?
8. What is More-than-Moore's law?
9. What is a silicon interposer technology?
10. Write a short note on 2.5D ICs.

Long Questions

1. Briefly explain the conventional packaging techniques and their drawbacks.
2. Discuss the recent advancement in packaging technology and explain how More-than-Moore's law can be achieved.
3. Discuss the silicon interposer technology in detail.
4. Discuss TSVs in detail and explain how they are different from conventional techniques.
5. What are the factors required for a technology consideration? Compare the 2.5D, 3D TSV, and 3D hybrid ICs on the basis of these factors.
6. What are the different methodologies for the realization of 3D IC?
7. Explain the following integration packaging techniques:
   a. 3D integrated microprocessor
   b. SRAM array integration
   c. NOC architecture

References

1. Papanikolaou, A., Soudris, D., and Radojcic, R. 2010. *Three Dimensional System Integration: IC Stacking Process an Design*. New York: Springer.
2. Kumar, V. R., Majumder, M. K., and Kaushik, B. K. 2014. Graphene based on-chip interconnects and TSVs—Prospects and challenges. *IEEE Nanotechnology Magazine* 8(4):14–20.
3. Majumder, M. K., Kumari, A., Kaushik, B. K., and Manhas, S. K. 2014. Signal integrity analysis in carbon nanotube based through-silicon via. *Active and Passive Electronic Components* 2014, Article ID 524107:1–7.
4. Lau, J. 2015. *3D IC Integration and Packaging*. New York: McGraw-Hill.

5. Kaushik, B. K., Majumder, M. K., and Kumari, A. 2014. *Fabrication and Modelling of Copper and Carbon Nanotube Based Through-Silicon Via, Design of 3D Integrated Circuits and Systems*. CRC Press/Taylor & Francis Group, Boca Raton, FL.

6. Majumder, M. K., Kumari, A., Kaushik, B. K., and Manhas, S. K. June 2014. Analysis of crosstalk delay using mixed CNT bundle based through silicon vias. In *Proceedings of IEEE Radio Frequency Integrated Circuits Symposium*, pp. 441–444. Tampa Bay, FL.

7. Kaushik, B. K., Majumder, M. K., and Kumar, V. R. 2014. Carbon nanotube based 3-D interconnects—A reality or a distant dream. *IEEE Circuits and Systems Magazine* 14(4):16–35.

8. Loh, G. H., and Xie, Y. 2010. 3D Stacked microprocessor: Are we there yet? *IEEE Micro* 30(3):60–64.

9. Yang, D. C., Xie, J., Swaminathan, M., Wei, X. C., and Li, E. P. 2013. A rigorous model for through-silicon vias with Ohmic contact in silicon interposer. *IEEE Microwave and Wireless Components Letters* 23(8):385–387.

10. Xie, B., and Swaminathan, M. 2014. FDFD modeling of signal paths with TSVs in silicon interposer. *IEEE* Transactions *on Components, Packaging and Manufacturing* 4(4):708–717.

11. Xie, B., Swaminathan, M., and Han, K. J. 2015. FDFD nonconformal domain decomposition method for the electromagnetic modeling of interconnections in silicon interposer. *IEEE Transactions on Electromagnetic Compatibility* 57(3):496–504.

12. Li, J., Wei, X., Wang, X., and Yu, H. 2014. Double-shielded interposer with highly doped layers for high-speed signal propagation. *IEEE Transactions on Electromagnetic Compatibility* 56(5):1210–1217.

13. Morrow, P. R., Park, C. M., Ramanathan, S., Kobrinsky, M. J., and Harmes, M. 2006. Three-dimensional wafer stacking via Cu-Cu bonding integrated with 65-nm strained-Si/low-k CMOS technology. *IEEE Electron Device Letters* 27(5):335–337.

14. Jacob, P., Erdogan, O., Zia, A., Belemjian, P. M., Kraft, R. P., and McDonald, J. F. 2005. Predicting the performance of a 3D processor-memory chip stack. *IEEE Design and Test of Computer* 22(6):540–547.

15. Alam, S. M., Troxel, D., and Thompson, C. V. 2002. A comprehensive layout methodology and layout-specific circuit analyses for three-dimensional integrated circuits. In *Proceedings of IEEE International Symposium on Quality Electronic Design*, pp. 246–251. San Jose, CA.

16. Yang, Y., Yu, M., Fang, Q., Song, J., Ding, L., and Lo, G.-Q. 2013. Through-Si-via (TSV) Keep-Out-Zone (KOZ) in SOI photonics interposer: A study of the impact of TSV-induced stress on si ring resonators. *IEEE Photonics Journal* 5(6):2700611(1)–2700611(12).

17. Kwon, W.-S. 2014. Cost effective, high performance 28 nm FPGA with new disruptive silicon less interconnect technology (SLIT). *IMAPS—International Microelectronics and Packaging Society and the Microelectronics Foundation* 1:599–605.

18. Guo, Y. X., Yin, W. Y., Sun, L., and Zhao, W. S. 2014. Electrothermal modelling and characterisation of submicron through-silicon carbon nanotube bundle vias for three-dimensional ICs. *IEEE Micro and Nano Letters* 9(2):123–126.

19. Xu, Z. 2011. Electrical evaluation and modeling of through-strata-vias (TSVs) in three-dimensional (3D) integration. PhD dissertation. Troy, NY: Rensselaer Polytechnic Institute.

20. Xie, Y. 2011. Three dimensional system integration: IC stacking process and design. In Papanikolaou, A., Soudris, D., and Radojcic, R. (eds.), *Microprocessor Design Using 3D Integration Technology*, pp. 211–236. Springer, New York.
21. Loh, G. H., Xie, Y., and Black, B. 2007. Processor design in 3D die-stacking technologies. *IEEE Micro* 27(3):31–48.
22. Kim, C., Burger, D., and Keckler, S. W. 2002. An adaptive, non-uniform cache structure for wire-delay dominated on-chip caches. In *Proceedings of the 10th International Conference on Architectural Support Programming Languages Operating Systems*, pp. 211–222. ACM, New York.
23. Lee, Y., Bang, S., Lee, I. et al. 2013. A modular 1 mm^3 die-stacked sensing platform with low power I^{2}C inter-die communication and multi-modal energy harvesting. *IEEE Journal Solid State Circuits* 48(1):229–243.

2

Through Silicon Vias: Materials, Properties, and Fabrication

2.1 Introduction

The importance of through silicon vias (TSVs) in the three-dimensional integrated circuit (3D IC) design has been discussed. The numerous advantages such as higher integration density, bandwidth, performance, and functionality obtained from 3D ICs would not have been possible without the application of TSVs. The performance of a 3D IC is primarily dependent on the choice of filler materials used in TSVs. Copper (Cu) is the most commonly used filler material in 3D TSVs. However, in recent years, Cu has faced certain challenges due to the fabrication limitations in achieving proper physical vapor deposition (PVD), seed layer deposition, and performance limitations due to electromigration and higher resistivity [1]. The resistivity can be attributed to the combined effects of scattering and the presence of highly diffusive barrier layer that increases the difficulty in obtaining a high aspect ratio via. Therefore, researchers are forced to find replacements to the Cu-based TSVs. Graphene-based materials such as carbon nanotubes (CNTs) and graphene nanoribbons (GNRs) have emerged as an interesting choice of filler materials due to their lower thermal expansion, Joule heating, and electromigration. Moreover, their higher current-carrying capability, long ballistic transport length, higher thermal conductivity, and mechanical strength provide exciting prospects for their application as filler materials in TSVs. Keeping in view the extraordinary properties demonstrated by graphene-based materials, it is believed that the ICs will soon contain graphene-based filler materials in TSVs in order to carry forward the "More-than-Moore" technologies.

The TSVs in 3D ICs carry the drawback of increasing the complexity of the overall system. This complicates the task of not only design engineers but also the fabrication engineers. The designing and fabrication goes hand in hand to successfully implement a realistic 3D IC. In spite of the numerous advantages believed to be provided by the TSV technology, there should be high-quality fabrication methods available in order to provide a minimum gap between the theoretical and experimental results. Therefore, it is important to discuss not only the advantages and properties but also the fabrication perspective of TSVs and the challenges faced during their fabrication.

2.2 History of Graphene

Graphene can be commonly referred to as a single layer of carbon atoms that have been obtained from graphite [2]. One of the earliest works connected with the emergence of graphene dates back to the 1840s when Schafhaeutl, a German scientist, reported the exfoliation and intercalation of graphite, resulting in graphite intercalation compounds [3,4]. The highly lamellar structure of graphite oxide was discovered by a British chemist, D. C. Brodie, in 1859. Brodie's approach to characterize the molecular weight of graphite may be recognized as modifications of methods used by Schafhaeutl [5,6]. Nearly 40 years later, a slightly different approach toward the preparation of graphite oxide through the addition of chlorate salt over the course of reaction instead as a single step was reported by Staudenmaier [7]. His work was regarded as one of the first examples toward delaminating graphite. In 1918, a detailed study on the properties of graphite oxide was carried out by Kohlschütter and Haenni [8]. A Canadian physicist, Philip Russel Wallace, had first discussed monolayer graphite in 1946, which was later termed as graphene [9].

Wallace was interested in studying the properties of graphite and believed that it was not possible for graphene to exist. In 1948, with the advancement in technology, Ruess and Vogt were the first to publish the images of few-layer graphite with the help of transmission electron microscopy (TEM) [10]. A German chemist, Hanns Peter Boehm, and his team were able to report single- and multilayer graphite flakes in 1962 [11]. Later, in 1968, Morgan and Somorjai were able to produce low-energy electron diffraction patterns by adsorption of small molecules onto Pt (100) [12]. A year later, the data produced in [12] was elucidated by scientist May to fabricate the monolayer graphite on the surface of Pt [13].

Over the following years, similar details of occurrence of single-layer graphite were reported; for example, monolayer graphite on the surface of Ni(100) was reported by Blakeley in 1970 [14]. With the surge in publications on monolayer graphite materials, Boehm and his team recommended the term "graphene" for single-layer graphite materials in 1986 [15–17]. After almost 11 years, these recommendations were acknowledged by the International Union of Pure and Applied Chemistry that included the definition of graphene into *Compendium of Chemical Terminology* [18]. In 1999, a micromechanical-based method was implemented by Ruoff and his team to produce thin lamellae consisting of graphene layers [19,20]. However, the process was unable to produce monolayer graphene. In 2002, a patent was filed by Nanotek Instruments, Inc. (Dayton, Ohio), for the production of graphene [21]. Two years later, in 2004, Geim and Novoselov were able to isolate graphene from graphite using a simple mechanical method [22]. They used a scotch tape to remove flakes of graphite from a graphite cube, until graphite flakes of only one atom thickness were obtained. This technique is popularly known as the "Scotch-tape" technique.

After 2004, there was a tremendous increase in research on graphene. In 2005, it was demonstrated by the researchers in Columbia University, New York, and the University of Manchester, Manchester, that graphene had quasiparticles which were massless Dirac fermions. Since the last decade, many leading organizations and research institutes such as International Business Machines (IBM), Samsung, University of California, Los Angeles (UCLA), and the University of Manchester have advocated millions of dollars to carry out the research on graphene; explore its unique electrical, mechanical, and optical properties; and diversify its applications in various domains. Considering the remarkable contributions of Geim and Novoselov toward the realization of potential graphene, they were awarded the Nobel Prize in Physics in 2010.

2.3 Carbon Nanotubes

Until the mid-1980s, diamond and graphite were the only two known forms of carbon allotropes. Allotropes can be defined as the materials that are made from the same basic element, but have different physical structures due to different bonding between atoms of the element. In 1985, Kroto et al. [23] were able to synthesize the new allotrope of carbon, C_{60}. They used a high pulse of laser light to vaporize a sample of graphite. The vaporized graphite was sent to mass spectrometer with the help of helium gas. The mass spectrometer detected the presence of C_{60}, a molecule consisting of 60 carbon atoms. The C_{60} had the shape of a soccer ball with 12 pentagon faces and 20 hexagonal faces as shown in Figure 2.1. The easy synthesis of C_{60} led

FIGURE 2.1
The structure of the bucky ball, C_{60}.

the group to propose the existence of another allotrope of carbon named as "bucky ball" due to its soccer ball-shaped structure. Soon after, other similar molecules consisting of only carbon such as C_{36}, C_{70}, and C_{76} were discovered. These new spherical-shaped, structured, carbon-based molecules were termed as fullerenes. The shape of the new allotrope of carbon did not end at the soccer ball-shaped structures, and long cylindrical tubelike structures were also reported, which are now known as CNTs.

2.3.1 Basic Structure of CNT

The structure of CNTs can be considered as a sheet of graphene rolled into a cylindrical structure. CNTs have fascinated the research world due to their extraordinary physical, electrical, and chemical properties. Many of the properties defy the conventional trends, and scientists are still discovering their unique properties and constantly making efforts to understand and explain the phenomenon for such distinctive behavior. One of the remarkable physical properties of CNT is the ability to scale down its thickness to a single atomic layer. This level of thickness is equivalent to about 1/50,000th times the thickness of a human hair. Another interesting physical property observed in CNTs is when two slightly different structured CNTs are joined together; the resultant junction formed can be used as an electronic device. The properties of the device formed are dependent on the type of CNTs used for their formation. The research world is still working to provide a practical solution for large-scale production of similar CNTs as it requires maintaining a high level of purity. The behavior of a bundle of CNTs can be easily studied and predicted once the bundle contains physically and chemically similar CNTs. The CNTs have intrigued the research world, and currently about five papers are published per day with CNTs as the central theme. This active involvement of the scientists in this material demonstrates how valuable and highly competitive this topic is [24].

2.3.1.1 Basic Structure of Single-Walled CNT

A single-walled CNT (SWCNT) can be assumed to be a structure formed when a single graphene sheet is rolled into a cylindrical structure (Figure 2.2). Depending on the shape of the circumference, CNTs can be classified as armchair (ac), zigzag (zz), and chiral CNTs as shown in Figure 2.2a, b, and c, respectively. The terms "zigzag" and "armchair" are inspired from the pattern in which the carbon atoms are arranged at the edge of the nanotube cross section. Graphene consists of sp^2-hybridized atoms of carbon that are arranged in a hexagonal pattern. The hexagonal carbon rings should join coherently when placed in contact with adjacent carbon atoms. Accordingly, in an SWCNT tube, all the carbon atoms (except at the edges) form hexagonal rings and are therefore equally spaced from one another. Recently, Xu et al. [25] have reported the fabrication of vertically grown CNT bundles with an

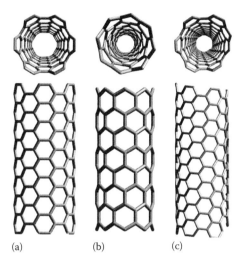

FIGURE 2.2
Sketches of three different SWCNT structures: (a) armchair, (b) zigzag, and (c) chiral.

average diameter of 50 μm and a pitch of 110 μm. The top view of the scanning electron microscope (SEM) image of the CNT bundles grown in two bonded wafers is shown in Figure 2.3a [25]. The detailed view of vertically grown two CNT bundles is shown in Figure 2.3b. It can be seen that the CNTs are well aligned. A high-magnification SEM image of the CNT bundle is shown in Figure 2.3c [25].

In spite of the hexagonal aromatic rings, SWCNT is considered more reactive than planar graphene. It is due to the fact that the hybridization in SWCNTs is not purely sp^2 and some degree of sp^3 hybridization is also present. It has been observed that with the decrease in SWCNT diameter, the degree of sp^3 hybridization increases [24]. This phenomenon causes variable overlapping of energy bands, which results in SWCNTs having versatile and unique electrical properties. It is studied that beyond the diameter of ~2.5 nm, the SWCNT tube collapses into a two-layer ribbon [26]. Moreover, a CNT with smaller diameter results in higher stress on the structure, although SWCNTs of ~0.4 nm diameter have been produced [27]. It is therefore natural to consider that a diameter of ~1 nm is the most suitable value with regard to energy consideration of SWCNTs. Encouragingly, there are no such restrictions on the length of the SWCNTs. The length is dependent on the processes and methods used for synthesis of the SWCNTs. SWCNTs of length ranging from micrometer to millimeter can be commonly found. Considering the diameter and length of an SWCNT, it is easy to intuitively conclude that SWCNT structures have exceptionally high aspect ratios.

A graphene sheet can be rolled in a number of different ways (see Figure 2.2). The mathematical expression that can be used to represent the various ways of rolling graphene into tubes is given as follows (see Figure 2.4):

$$PX = C_h = pb_1 + qb_2 \tag{2.1}$$

where:

C_h and θ are the chirality vector and the chirality angle, respectively
p and q are integers

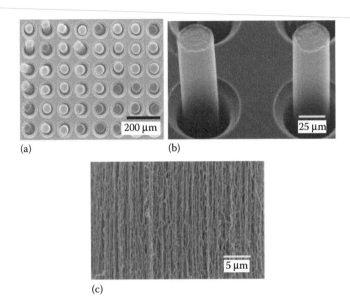

FIGURE 2.3
SEM images for CNTs grown out of the top bonded wafer: (a) overview of CNT bundles, (b) detailed view of two CNT bundles taken with an angle of 45°, and (c) a high-magnification SEM image of the CNT bundle. (Reproduced with permission from Xu, T. et al., *Applied Physics Letters*, 91, 042108-1–042108-3, 2007.)

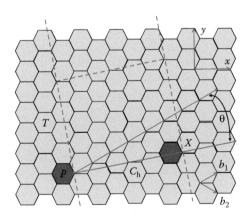

FIGURE 2.4
Sketch representing the procedure to obtain an SWCNT, starting from a graphene sheet.

The unit vectors b_1 and b_2 are defined as

$$b_1 = \frac{b\sqrt{3}}{2}x + \frac{b}{2}y \quad \text{and} \quad b_2 = \frac{b\sqrt{3}}{2}x - \frac{b}{2}y \tag{2.2}$$

where:
$b = 2.46$ Å

$$\cos\theta = \frac{2p+q}{2\sqrt{p^2+q^2+pq}}$$

The vector PX is normal to the tube axis and θ is considered with respect to the zigzag axis: C_h that results in a zz type of CNTs (zz-type CNTs are discussed later). Moreover, the diameter, D, of a nanotube is dependent on the C_h by the following relation:

$$D = \frac{|C_h|}{\pi} = \frac{b_{C=C}\sqrt{3\left(p^2+q^2+pq\right)}}{\pi} \tag{2.3}$$

where 1.41Å(graphene) $\leq b_{C=C} \leq 1.44$Å(C_{60}).

The C=C bond length in the hexagonal ring structure of SWCNT slightly increases from the C=C bond length in graphene due to the curved structure of SWCNT. The degree of curvature in an SWCNT cannot exceed that in C_{60} molecule, resulting in the upper limit of the C=C bond length in SWCNTs. Similarly, the degree of curvature in an SWCNT cannot be less than that in a flat graphene structure, resulting in the lower limit of C=C bond length in SWCNTs. Moreover, it can be observed that C_h, θ, and D can be expressed in terms of p and q. Because SWCNTs can be identified by C_h, θ, and D values, it is sufficient to define SWCNTs through p and q values by denoting them as (p, q). The p and q values for a particular SWCNT can be easily obtained by counting the number of hexagonal rings separating the margins of C_h vector following b_1 first and then b_2 [28]. Based on the (p, q) representation, zz SWCNTs can be denoted as $(\bar{p}, 0)$ having $\theta = 0°$; ac SWCNTs can be denoted as $(\bar{p}, \bar{p})$ having $\theta = 30°$; chiral SWCNTs can be denoted as (p, q) having $0 < \theta < 30°$.

From Figure 2.4, it can be observed that having C_h direction perpendicular to any carbon bond directions results in zz SWCNT ($\theta = 0°$), whereas having C_h direction parallel to any carbon bond directions will result in ac SWCNT ($\theta = 30°$). In chiral SWCNTs, $0 < \theta < 30°$ due to hexagonal rings in a graphene sheet.

2.3.1.2 Basic Structure of Multiwalled CNT

A multiwalled CNT (MWCNT) is a bit more intricate in structure compared to an SWCNT. Unlike a single graphene shell in SWCNT, there are multiple graphene shells in an MWCNT. The MWCNTs have two or more numbers of CNT shells that are concentrically rolled up. The structure of the MWCNT between the two contacts is shown in Figure 2.5, in which the inset figure shows its cross-sectional view. The intershells are separated by the van der Waals gap, $\delta \sim 0.34$ nm. The diameter of the outermost CNT shell can be varied from nanometers to several tens of nanometers. The current-carrying capabilities of the MWCNT are similar to the SWCNT bundle; however, the MWCNTs are easier to fabricate. The diameters of the outermost and innermost shells are denoted by D_N and D_1, respectively. The ratio of D_1/D_N varies in different MWCNTs; the values between 0.3 and 0.8 have been observed in [29–31]. The density of 10^6 cm^{-2} has been obtained in [29] with a D_1/D_N of 0.5. Close et al. [32] reported the fabrication of MWCNTs with 80 shells, as shown in Figure 2.6. The authors proposed a versatile method that is ideally suited for fabricating MWCNT TSVs with extensive electrical properties.

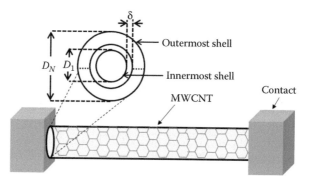

FIGURE 2.5
The structure of MWCNT placed between the two contacts.

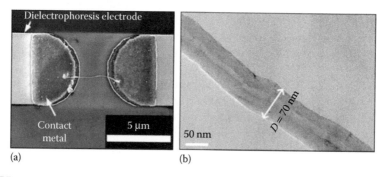

FIGURE 2.6
(a) The fabricated structure of an MWCNT between two metal contacts and (b) a close-up view of an MWCNT consisting of 80 shells. (Reproduced with permission from Close G. F. and Wong, H. S. P. *IEEE Transactions Nanotechnology, 7,* 596–600, 2008.)

2.3.2 Semiconducting and Metallic CNTs

CNTs can act as semiconducting or metallic based on the pattern of the CNT circumference. The ac CNTs always act as metallic, whereas the zz CNTs act as either metallic or semiconducting depending on the chiral indices. This section presents the behavior of zz CNTs and their dual nature.

Because CNT is a rolled-up sheet of graphene, an appropriate boundary condition is required to explore the band structure. If CNT can be considered as an infinitely long cylinder, there are two wave vectors associated with it: (1) the wave vector parallel to CNT axis $k_{||}$ that is continuous in nature due to the infinitely long length of CNTs and (2) the perpendicular wave vector $k_\perp$ that is along the circumference of the CNT. These two wave vectors must satisfy a periodic boundary condition (i.e., the wave function repeats itself as it rotates 2π around a CNT) [33]:

$$k_\perp \cdot C_h = \pi D k_\perp = 2\pi m \tag{2.4}$$

where m is an integer.

The quantized values of allowed $k_\perp$ for CNTs are obtained from the boundary condition. The cross-sectional cutting of the energy dispersion with the allowed $k_\perp$ states results in one-dimensional (1D) band structure of graphene as shown in Figure 2.7a. This is called zone-folding scheme of obtaining the band structure of CNTs. Each cross-sectional cutting gives rise to 1D sub-band. The spacing between the allowed $k_\perp$ states and their angles with

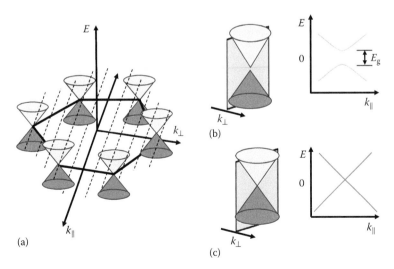

FIGURE 2.7
(a) A first Brillouin zone of graphene with conic energy dispersions at six K points. The allowed $k_\perp$ states in CNT are presented by dashed lines. The band structure of the CNT is obtained by cross sections as indicated. Zoom ups of the energy dispersion near one of the K points are schematically shown along with the cross sections by allowed $k_\perp$ states and resulting 1D energy dispersions for (b) a semiconducting CNT and (c) a metallic CNT.

respect to the surface Brillouin zone determine the 1D band structures of CNTs. The band structure near the Fermi level is determined by the allowed $k\perp$ states that are closest to the K points. When the allowed $k_\perp$ states pass directly through the K points as shown in Figure 2.7c, the energy dispersion has two linear bands crossing at the Fermi level without a bandgap. However, if the allowed $k_\perp$ states miss the K points as shown in Figure 2.7b, there would be two parabolic 1D bands with an energy bandgap. Therefore, two different kinds of CNTs can be expected depending on the wrapping indices: first, the semiconducting CNTs with bandgap as in Figure 2.7b, and second, the metallic CNTs without bandgap as in Figure 2.7c.

Using the approach of 1D sub-bands discussed in Section 2.3.1.1, the 1D sub-band closest to the K points for zigzag CNTs is investigated here. The zigzag CNTs can be either metallic or semiconducting depending on their chiral indices. Because the circumference is $\bar{p}b$ ($C_h = \bar{p}b_1$), the boundary condition in Equation 2.4 becomes [33]

$$k_x \bar{p}b = 2\pi m \tag{2.5}$$

There is an allowed k_x that coincides with the K point at $(0, 4\pi/3b)$. This condition arises when $\bar{p}$ has a value in multiple of 3 ($\bar{p} = 3\bar{q}$, where $\bar{q}$ is an integer), Therefore, substituting in Equation 2.5 [33],

$$k_x = \frac{2\pi m}{\bar{p}b} = \frac{3Km}{2\bar{p}} = \frac{Km}{2\bar{q}} \tag{2.6}$$

There is always an integer m ($= 2\bar{q}$) that makes k_x pass through K points, and these kinds of CNTs (with $\bar{p} = 3\bar{q}$) are always metallic without bandgap as shown in Figure 2.7b. There are two cases when p is not a multiple of 3. If $\bar{p} = 3\bar{q}+1$, the k_x is closest to the K point at $m = 2\bar{q}+1$ (as in Figure 2.7c).

$$k_x = \frac{2\pi m}{\bar{p}b} = \frac{3Km}{2\bar{p}} = \frac{3K(2\bar{q}-1)}{2(3\bar{q}+1)} = K + \frac{K}{2}\frac{1}{3q+1} \tag{2.7}$$

Similarly, for $\bar{p} = 3\bar{q}-1$, the allowed k_x closest to K is when $m = 2\bar{q}-1$; hence,

$$k_x = \frac{2\pi m}{\bar{p}b} = \frac{3Km}{2\bar{p}} = \frac{3K(2\bar{q}-1)}{2(3\bar{q}-1)} = K - \frac{K}{2}\frac{1}{3\bar{q}-1} \tag{2.8}$$

In these two cases, allowed k_x misses the K point by

$$\Delta k_x = \frac{K}{2}\frac{1}{3\bar{q}\pm 1} = \frac{2}{3}\frac{\pi}{\bar{p}b} = \frac{2}{3}\frac{\pi}{\pi D} = \frac{2}{3D} \tag{2.9}$$

From the above equation, it is inferred that the smallest misalignment between an allowed k_x and a K point is inversely proportional to the diameter.

Thus, from the slope of a cone near *K* points (Equation 2.4), the bandgap E_g can be expressed as

$$E_g = 2 \times \left(\frac{\partial E}{\partial k}\right) \times \frac{2}{3D} = 2\hbar v_F \left(\frac{2}{3D}\right) \approx 0.7 \text{eV}/D \text{ (nm)} \tag{2.10}$$

Therefore, semiconducting CNTs (D = 0.8–3 nm) exhibit bandgap ranging from 0.9 to 0.2 eV. Depending on the value of $\bar{p}$, where $\bar{p}$ is the remainder when *p* and *q* are divided by 3, SWCNTs (represented by (*p*, *q*)) can be of three types:

$\bar{p} = 0$; metallic with linear sub-bands crossing at *K* points

$\bar{p} = 1, 2$; semiconducting with a bandgap, $E_g \sim 0.7 \text{eV}/D$ (nm)

Similar treatment can also be applied for armchair CNTs ($\bar{p}, \bar{p}$), arriving at the conclusion that they are always metallic.

2.3.3 Properties and Characteristics

The atomic arrangements of carbon atoms are responsible for the unique electrical, thermal, and mechanical properties of CNTs [34,35]. The sp^2 bonding delivers the high conductivity and mechanical strengths to the CNTs. The unique properties of CNTs are discussed in Sections 2.3.3.1 through 2.3.3.6.

2.3.3.1 Strength and Elasticity

Due to the sp^2 hybridization, each carbon atom in a single sheet of graphite is connected via strong sigma bonds to three neighboring atoms. Thus, CNTs can exhibit the strongest basal plane elastic modulus and hence are expected to be the ultimate high-strength fiber. The elastic modulus of SWCNTs is much higher than steel, which makes them highly resistant. Although pressing on the tip of nanotube will cause it to bend, the nanotube returns to its original state as soon as the force is removed. This property makes CNTs extremely useful as probe tips for high-resolution scanning probe microscopy. Although the current Young's modulus of SWCNTs is about 1 Tpa, a much higher value of 1.8 Tpa has also been reported [36]. For different experimental measurement techniques, the values of Young's modulus vary in the range of 1.22–1.26 Tpa depending on the size and chirality of the SWNTs [35]. It has been observed that the elastic modulus of CNTs is not strongly dependent on the diameter. Primarily, the moduli of CNTs correlate to the amount of disorder in the nanotube walls [37].

2.3.3.2 Thermal Conductivity and Expansion

CNTs can exhibit superconductivity below 20K (approximately –253°C) due to the strong in-plane sigma bonds in between carbon atoms. The sigma bond provides exceptional strength and stiffness against axial strains. Moreover, the larger interplane and zero in-plane thermal expansion of SWCNTs results in high flexibility against nonaxial strains.

Due to their high thermal conductivity and large in-plane expansion, CNTs exhibit exciting prospects in nanoscale molecular electronics, sensing and actuating devices, reinforcing additive fibers in functional composite materials, and so on. Recent experimental measurements suggest that the CNT-embedded matrices are stronger compared to bare polymer matrices [38]. Therefore, it is expected that the nanotube may also significantly improve the thermomechanical and thermal properties of the composite materials.

2.3.3.3 Field Emission

Under the application of strong electric field, tunneling of electrons from the metal tip to vacuum results in the phenomenon of field emission. Field emission results from the high aspect ratio and small diameter of CNTs. The field emitters are suitable for the application in flat panel displays. For MWCNTs, the field emission properties occur due to the emission of electrons and light. Without an applied potential, the luminescence and light emission occurs through the electron field emission and visible part of the spectrum, respectively.

2.3.3.4 Aspect Ratio

One of the exciting properties of CNTs is the high aspect ratio, which infers that a lower CNT load is required compared to other conductive additives to achieve similar electrical conductivity. The high aspect ratio provides unique electrical conductivity in CNTs compared to the conventional additive materials such as chopped carbon fiber, carbon black, or stainless steel fiber.

2.3.3.5 Absorbent

CNTs and CNT composites have been emerging as perspective absorbing materials due to their light weight, larger flexibility, high mechanical strength, and large surface area. Therefore, CNTs emerge out as an ideal candidate for use in gas, air, and water filtration. The absorption frequency ranges of SWCNT–polyurethane composites broaden from 6.4–8.2 (1.8 GHz) to 7.5–10.1 (2.6 GHz) to 12.0–15.1 GHz (3.1 GHz) [39]. A lot of research has already been carried out for replacing the activated charcoal with CNTs for certain ultra-high purity applications.

2.3.3.6 Conductivity

CNTs are assumed to be the most electrically conductive materials. However, it is quite difficult to control the chirality of the SWCNT shells, and therefore, statistically only one-third of the CNTs in a bundle are assumed to be conducting and the rest of them are semiconducting. However, because of large diameters, the shells of MWCNTs would be conductive even if they have semiconductor characteristics. This fact can be realized by the following explanation. The energy gap between the conduction band edge and the Fermi level of a CNT shell is defined as follows:

$$E_g = \frac{v_0 p_{C=C}}{d}$$ (2.11)

where:
d is the CNT diameter
v_0 is the nearest-neighbor tight-binding (TB) parameter
$p_{C=C}$ is the nearest neighbor C=C bond length of ~ 0.142 nm

From the above equation, it can be observed that the bandgap is inversely proportional to its diameter. For a semiconducting CNT shell, whose diameter is 20 nm, the bandgap is observed to be less than 0.0258 eV, which can be smeared by the environmental temperature. Therefore, the semiconducting CNT shells are conductive if the diameter is greater than 20 nm. The detailed conductivity comparison between MWCNTs and SWCNTs will be discussed in Sections 2.3.3.6.1 and 2.3.3.6.2.

2.3.3.6.1 Conductivity Comparison

The performance of TSV primarily depends on the conductivity of the TSV filler material. The conductivity comparison among Cu, SWCNT, and MWCNT is analyzed in this section.

2.3.3.6.1.1 SWCNT Conductivity The conductivity of SWCNT [40] is expressed as follows:

$$\sigma_{SWCNT} = \frac{4G_0 l_0 D F_m}{\sqrt{3}(D+\delta)^2} \frac{l}{(l+l_0 D)}$$ (2.12)

where:
l, D, F_m are the TSV height, shell diameter, and fraction of metallic CNTs in the bundle, respectively
l_0 is 10^3
δ is 0.34 nm
G_0 is the quantum conductance, equal to $2e^2/h$ (where e is the charge of an electron and h is the Planck constant)

For $l > l_0 D$, the above equation can be expressed as

$$\sigma_{SWCNT} \approx \frac{4G_0 l_0 D F_m}{\sqrt{3}(D+\delta)^2}$$ (2.13)

From the above equation, it can be observed that for longer TSVs, the conductivity of SWCNT is independent of length.

2.3.3.6.1.2 MWCNT Conductivity   The conductivity of MWCNT [41] is expressed as

$$\sigma_{MWCNT} = \frac{G_0 l}{2\delta} \left\{ \left(1 - \frac{D_{min}^2}{D_{max}^2}\right)\frac{a}{2} + \left(b - \frac{l}{l_0}a\right) \times \left[\left(\frac{1}{D_{max}} - \frac{D_{min}}{D_{max}^2}\right) - \frac{l}{D_{max}^2 l_0} \ln\left(\frac{D_{max} + \frac{l}{l_0}}{D_{min} + \frac{l}{l_0}}\right) \right] \right\} \quad (2.14)$$

where:

D_{max} and D_{min} are the outermost and the innermost shell diameter of an MWCNT, respectively

a and b are constants and the values are 0.0612 nm^{-1} and 0.425, respectively [42]

From the above equation, it can be observed that for $l > (l_0 b/a)$, the conductivity increases with an increase in D_{max}.

The conductivity comparison plot among Cu, SWCNT bundles, and MWCNTs is shown in Figure 2.8. It can be observed that for shorter TSV lengths, the conductivity of SWCNT bundle is higher than that of MWCNT, whereas for longer lengths, MWCNTs can potentially have conductivities several times larger than copper or even SWCNTs, which is essential for TSV applications. It is worth noting that the best-case scenario was considered for SWCNTs in which they were densely packed so that the highest conductivity

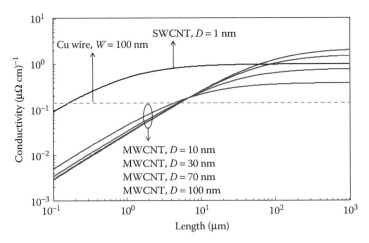

FIGURE 2.8
The conductivity comparison among Cu, SWCNT, and MWCNT.

is obtained. However, in contrast to this, an average-case scenario was considered for MWCNTs in which the innermost diameter is half of the outermost shell diameter. The innermost diameters were considered as 5, 15, 35, and 50 nm for their respective outer diameters of 10, 30, 70, and 100 nm. However, the best-case scenario for MWCNTs would have been when the innermost diameter had been 1 nm. But still, for longer TSVs, the performance of MWCNTs is better than that of SWCNTs.

2.4 Graphene Nanoribbons

In 1996, Fujita and his group provided a theoretical model of GNRs to observe the edge and nanoscale dimension effect in graphene [43,44]. Recent developments in GNRs have aroused a lot of research interests for their potential applications in the area of interconnects, TSVs, and field-effect transistors [45–47]. Ballistic transport in GNR makes it suitable not only for interconnects and TSVs but also for switching transistors. A monolithic system can be constructed using the single-layer GNR for both transistors and interconnects. For nanoscale device dimensions, Cu-based TSVs are mostly affected by grain boundaries and sidewall scattering. It has been predicted that GNRs will outperform the Cu-based TSVs for smaller widths [48]. In a high-quality GNR sheet, the mean free path (MFP) is ranging from 1 to 5 μm. GNRs can carry large current densities more than 10^8 A/cm^2. They also offer higher carrier mobility that can reach up to 3×10^3 cm^{2}v^{-1}s^{-1} [49].

2.4.1 Basic Structure of GNR

A GNR is a single sheet of graphene layer, which is extremely thin and limited in width, such that it results in a 1D structure [50]. As a result, GNRs can be considered as an unrolled version of CNTs with most of their electronic properties similar to those of CNTs. Depending on the termination of their width, GNRs can be divided into chiral and nonchiral GNRs. Chiral GNRs can be further classified as armchair (ac) and zigzag (zz) GNRs as shown in Figure 2.9a and b, respectively.

Please note that the terms "armchair" and "zigzag" are used for both GNRs and CNTs. However, confusingly, these nomenclatures are used in opposite ways. For GNRs, these terms indicate the pattern of the GNR edge, whereas for CNTs, they indicate the CNT circumference. Therefore, the unrolled armchair CNT can be visualized as a zigzag GNR and vice versa.

Depending on the stacked graphene sheets, GNRs are classified as single-layer GNRs (SLGNRs) and multilayer GNRs (MLGNRs). The geometry of an MLGNR is shown in Figure 2.10. The MLGNR TSV consists of N number of layers, with an interlayer distance δ, a width w, and a thickness t.

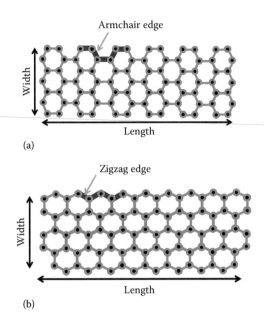

FIGURE 2.9
The structure of GNR: (a) armchair and (b) zigzag.

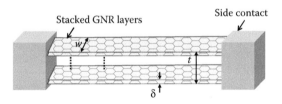

FIGURE 2.10
The geometric structure of an MLGNR TSV.

From the fabrication point of view, it is evident that the growth of the GNR can be more easily controlled than that of the CNT because of its planar structure. This makes them compatible with the conventional lithography techniques [51]. Using the electron beam lithography technique, Murali et al. [52] fabricated an MLGNR with 10 layers; the SEM image is shown in Figure 2.11.

Moreover, higher electrical conductivity of MLGNR can be obtained either by enhancing the carrier mobility or by increasing the number of carriers. The carrier mobility can be increased by intercalation doping of arsenic pentafluoride (AsF_5) vapor. Using the AsF_5 doping, the conductivity of MLGNR can be increased up to 3.2×10^5 S/cm, which is almost 1.5 times higher than that of copper [53]. Additionally, the easier fabrication process of MLGNR makes it as a promising candidate for TSV materials [54].

FIGURE 2.11
The fabricated structure of an MLGNR consisting of 10 layers. (Reproduced with permission from Murali, R. et al., *IEEE Electron Device Letters*, 30, 611–613, 2009.)

2.4.2 Semiconducting and Metallic GNRs

GNRs can act as semiconducting or metallic based on the pattern of the GNR edge. The zigzag edge-patterned GNRs always act as metallic, whereas the armchair edge-patterned GNRs can act as either metallic or semiconducting depending on the number of carbon atoms present across the width of GNRs. This section presents the behavior of armchair GNRs and their dual nature.

The typical structure of an ac GNR is shown in Figure 2.9a, in which the number of carbon atoms N across its width is 7. For the metallic/semiconducting behavior of GNRs, it is necessary to understand the electronic band structures of GNRs. The band structures of GNRs are obtained using a TB model. Using the TB approach, the band structures of 23- and 24-atom-wide armchair GNRs are shown in Figure 2.12a and b, respectively. It can

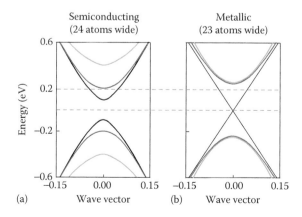

FIGURE 2.12
Band structures of (a) semiconducting and (b) metallic armchair GNRs whose widths are 6.02 nm and 5.78 nm, respectively.

be observed that the metallic GNR has zero bandgap, whereas the semiconducting GNR has 0.2 eV bandgap. The ac GNRs act as metallic if $N = 3a + 2$ and semiconducting if $N = 3a + 1$ or $3a + 3$, where a is an integer. The zz GNRs are always metallic, independent of the value of N [55].

2.4.3 Properties and Characteristics

GNRs can conduct much higher current than Cu due to the pi bonding of p_z orbitals. The thermal conductivity of GNR is also larger than that of Cu. Moreover, GNRs have large MFP that leads to large current conductance.

Most of the physical and electrical properties of GNRs are similar to those of CNTs. However, compared to CNTs, the growth of GNRs is considered to be more controllable due to their planar structure. Moreover, the major advantage of GNRs over CNTs is that both transistor and interconnect can be fabricated on the same continuous graphene layer, which, unlike CNTs, are free from Stone–Wales defects [56]. Therefore, one of the manufacturing difficulties regarding the formation of metal–nanotube contact can be avoided. Due to the lower resistivity, MLGNRs are often preferred over SLGNRs as a suitable TSV material. However, MLGNRs, fabricated till date, have displayed some level of edge roughness [51,57]. The electron scattering at the rough edges reduces the MFP, which substantially lowers the conductance of MLGNRs. This fundamental challenge limits the performance of MLGNR-based TSVs. The value of MFP primarily depends on the level of edge roughness. Section 2.4.3.1 discusses the effect of edge roughness on the MFP.

2.4.3.1 MFP of GNR

The effective MFP of GNR, λ_{eff}, depends on the scattering effects due to phonons, λ_{ph}, defects, λ_d, and edge roughness, λ_n. Using Matthiessen's rule, the λ_{eff} can be expressed as

$$\frac{1}{\lambda_{eff}} = \frac{1}{\lambda_d} + \frac{1}{\lambda_n} + \frac{1}{\lambda_{ph}} \tag{2.15}$$

For the TSV applications (low bias), the MFP corresponding to λ_{ph} is observed as extremely large, that is, tens of micrometers, and therefore, its effect can be neglected for the modeling of GNR scattering resistance [55]. Consequently, λ_d and λ_n dominate and determine the overall value of λ_{eff}.

2.4.3.1.1 MFP Variation for Smooth Edges

This section describes the procedure to calculate the MFP of GNR with smooth edges, that is, only considering scattering effects due to defects, λ_d.

According to the experimental measurements reported in [57], the MFP corresponding to λ_d is about 1 μm for a single layer GNR, which is width independent. However, in multilayer GNR, the MFP reduces due to the inter-sheet electron hopping [58,59]. The λ_d of MLGNR can be extracted by measuring the in-plane conductivity of GNR. Using the in-plane conductivity of $G_{sheet} = 0.026$ (μΩ·cm)$^{-1}$ [59], the layer spacing of 0.34 nm, and $E_f = 0$ of a neutral MLGNR, the λ_d is extracted as 419 nm by solving the following equation [60]:

$$G_{sheet} = \frac{2q^2}{h} \cdot \frac{\pi \lambda_d}{hv_f} \cdot 2k_B T \ln\left[2\cosh\left(\frac{E_f}{2k_B T}\right)\right] \qquad (2.16)$$

To increase the conductivity of MLGNR, AsF$_5$ intercalated graphite is used. The in-plane conductivity $G_{sheet} = 0.63$ (μΩ·cm)$^{-1}$ and the carrier concentration $n_p = 4.6 \times 10^{20}$ cm^{-3} are observed for the AsF$_5$-intercalated graphite [61]. Using the simplified TB model, E_f can be expressed as

$$E_f = hv_f\left(\frac{n_p \cdot \delta}{4\pi}\right)^{1/2} \qquad (2.17)$$

where $\delta = 0.575$ nm is the average layer spacing between adjacent graphene layers. Using the expressions (2.15), (2.16), and (2.17), E_f and λ are expressed as 0.6 eV and 1.03 μm, respectively.

2.4.3.1.2 MFP Variation for Rough Edges

The MFP corresponding to diffusive scattering at the edges is a function of edge backscattering probability, P, and the average distance an electron travels along the length before hitting the edge. The MFP for the nth sub-band due to edge scattering can be expressed as [48]

$$\lambda_n = \frac{W}{P}\sqrt{\left(\frac{E_f / \Delta E}{n}\right)^2 - 1} \qquad (2.18)$$

where ΔE is the gap between the sub-bands.

The effective MFP for different values of edge roughness probabilities of MLGNR is shown in Figure 2.13. The MFP due to defects and Fermi level are assumed to be 419 nm and 0.2 eV, respectively. It is observed in Figure 2.13 that the edge roughness reduces the MFP by more than 1 order of magnitude, particularly for narrow widths. However, the reduction of MFP is highly dependent on the P value.

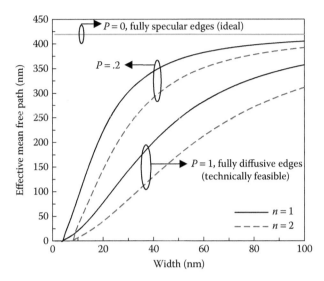

FIGURE 2.13
The MFP of MLGNR for the first two lowest sub-bands at different TSV widths. (Reproduced with permission from Kumar, V. R. et al., *IEEE Transactions on Nanotechnology*, 14, 484–492, 2015.)

2.5 Properties of TSVs

In scaled 3D technologies, the overall delay and reliability of an IC is primarily dependent on TSV self and coupled parasitics. It is necessary to reduce the electromagnetic interference, delay, crosstalk, and effective wire capacitance of TSVs. Moreover, due to the stacking of ICs on top of each other, thermal and stress management issues are very critical in 3D ICs [62]. It has therefore become essential to study and understand the electrical, thermal, mechanical, and thermomechanical properties of TSVs.

2.5.1 Electrical Properties

TSVs provide short signal paths that result in reduced resistive, inductive, and capacitive components. Moreover, the parasitics provided by the TSVs are lesser compared to the typical two-dimensional (2D) global interconnect wires. Due to these reduced parasitics in 3D TSV-based technologies, the signaling performance and driving capability of ICs are improved significantly. The TSV-based 3D IC can provide improved electrical characteristics to address the limits in leakage and electrical performance of complementary metal–oxide–semiconductor (CMOS) ICs beyond 11–16 nm [63]. Therefore, it is essential to understand the electrical properties of TSVs to comprehend the benefits of TSV-based systems.

The electrical properties of the TSVs are largely dependent on the capacitance and leakage current. Various system analyses, such as dynamic power dissipation, propagation delay, and crosstalk, all depend on the capacitance values of TSVs. The insulation capabilities of TSVs can be determined by the evaluation of the leakage current. By plotting the varying capacitance and leakage current values against the varying voltage values, the electrical properties of the TSVs can be characterized. A typical TSV setup is shown in Figure 2.14, in which the TSV metal is isolated from the Si substrate via the oxide layer [64]. The Si substrate is connected to zero potential and acts as a ground, whereas the TSV metal is connected to the signal source and acts as a gate. Therefore, a metal–oxide–semiconductor capacitor (MOSCAP) is formed among the TSV, the oxide liner, and the Si substrate. This cylindrical MOSCAP is further regarded as the TSV capacitance. The value of the TSV capacitance primarily depends on the depletion of the Si substrate.

The C–V plot obtained at 25°C, for a frequency range of 10 KHz–1 MHz, is shown in Figure 2.15. It can be observed that the C–V plot shows the evidence of separate accumulation, depletion, and inversion regions, similar to the C–V plot obtained for the MOS capacitor [65]. Due to the defects, the high-frequency (1 MHz) signals at the TSV metal cause disturbance of the minority carriers in the substrate, resulting in the kink in the C–V plot. Nevertheless, near the operating voltage regions of 0–5 V, the value of the TSV capacitance is close to the minimum depletion capacitance value, which is desirable for improved TSV performance. Moreover, it can be observed from Figure 2.15 that at 25°C, for all frequencies, the effect of carrier charges is not significant, as suggested by the hysteresis curve.

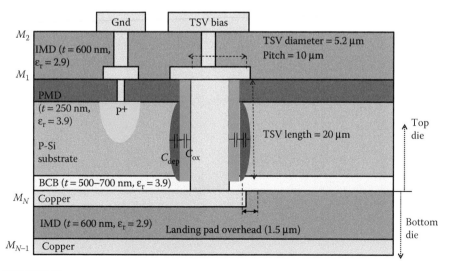

FIGURE 2.14
Schematic of TSV capacitance and leakage measurement.

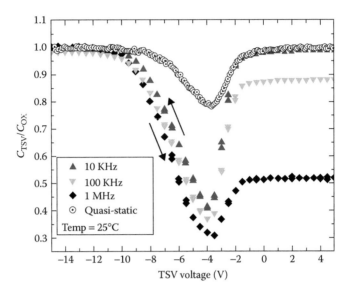

FIGURE 2.15
C–V characteristic of TSV for varying frequencies at 25°C. (From Katti, G. et al., *Proceedings of IEEE Conference on IITC*, pp. 1–3, Burlingame, CA, 2010. With Permission.)

Figure 2.16 shows the leakage I–V curve with change in temperature. The variation of the leakage current is nonlinear with voltage. Moreover, it can be observed that the leakage current increases with an increase in temperature. This is due to the fact that the leakage current comprises minority charge carriers, which in turn increase with increase in temperature. Even at temperature as high as 150°C, the maximum leakage current calculated

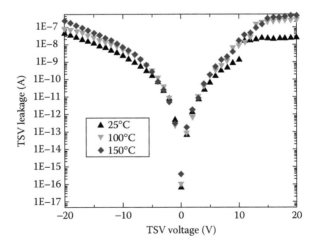

FIGURE 2.16
Leakage current variation in TSV with temperature. (From Katti, G. et al., *Proceedings of IEEE Conference on IITC*, pp. 1–3, Burlingame, CA, 2010. With Permission.)

is <1 µA, and thus, the isolation property of the TSV oxide is intact even at high temperatures.

To further enhance the performance of a TSV by reduction in the leakage current, the TSV oxide can be replaced by benzocyclobutene (BCB) liners as insulators [66]. Figure 2.17 shows the variation in the leakage current for a temperature range of 25°C–125°C, while using BCB liners instead of SiO_2 oxide. It can be observed that the leakage current in BCB liner-based TSVs at 125°C is at least 3 orders of magnitude less than that in the conventional SiO_2 liner-ased TSVs at 100°C [64]. In spite of better performance in terms of the reduced leakage current in BCB liner-based TSVs, the fabrication of SiO_2 liner-based TSVs is much easier than that of other liner-based TSVs. Hence, for practical applications, the fabrication point of view for different liner-based TSVs should also be considered.

2.5.2 Thermal Transport

The thermal properties of TSVs also play a significant role and are important for characterization of the system due to the following reasons [67]:

1. The power generated per unit area in 3D IC is quite high.

2. Due to limited space within adjacent TSVs, the cooling channels may not be properly placed. This may result in temperatures reaching extremely high values within the 3D IC.

3. On-chip hot spots may be created by thin chips of 3D IC.

4. The multifunctional chips of extremely reduced area may generate extreme heat flux.

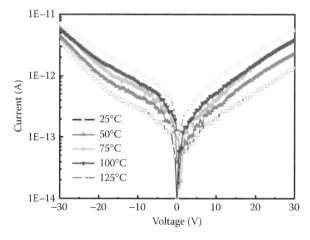

FIGURE 2.17
Leakage current variation in TSV with temperature from 25°C to 125°C. (From Huang, C. et al., *IEEE Transactions on Components, Packaging and Manufacturing Technology*, 4, 1936–1946, 2014. With Permission.)

The flow of current through TSV is responsible for Joule heating that causes a rise in temperature. Because the TSVs with lower aspect ratios have higher thermal resistance, the Joule heating establishes a lower limit constraint on the TSV aspect ratio. Moreover, as the barrier layer in a TSV is thermally insulating, it may significantly increase the temperature. Also, the temperature variation is primarily dependent on the barrier layer thickness. Moreover, the thermal gradients resulting from the creation of hot spots within a TSV may lower the reliability of the overall system and increase the leakage current.

For a typical TSV setup (see Figure 2.14), the C–V plot at 1 MHz frequency and at different temperature values is shown in Figure 2.18. It can be observed that the hysteresis curve is quite narrow till 100°C, whereas it is quite broad at 150°C. This is because the influence of mobile ions is insignificant below 100°C, whereas they become significant at higher temperatures. Moreover, it can be observed in Figure 2.18 that the TSV capacitance increases with an increase in temperature, at operating voltage regions of 0–5 V. This is because with the rise in temperature, the generation of minority charge carriers in the semiconductor increases, and these excess charge carriers respond to the high-frequency signal applied at the gate, thereby increasing the capacitance value. The variation of the leakage current with increase in temperature has already been discussed in Section 2.5.1.

A typical TSV along with Cu bonding is shown in Figure 2.19. The difference in the diameters of the TSV and the bonding results in the development of thermal strain. The value of the equivalent strain developed for different TSV diameters at fixed bonding diameter (30 μm) is shown in Figure 2.20a. It can be noticed that the smaller the TSV diameter, the lesser is the strain experienced by the TSV, whereas the smaller the diameter difference between the TSV and the bonding, the lesser is the strain experienced by the bonding.

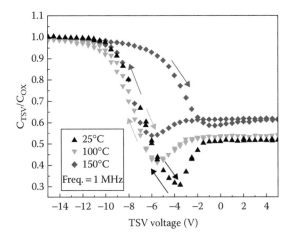

FIGURE 2.18
C–V characteristic of TSV for varying temperatures at 1 MHz. (From Katti, G. et al., *Proceedings of IEEE Conference on IITC*, pp. 1–3, Burlingame, CA, 2010. With Permission.)

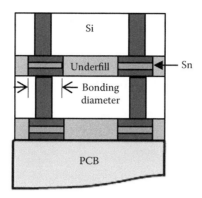

FIGURE 2.19
Schematic of Cu bonding and TSV.

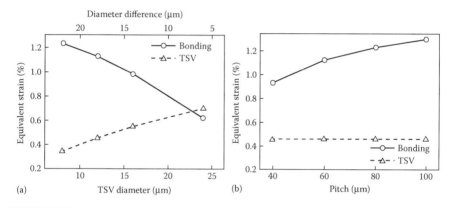

FIGURE 2.20
Change in equivalent strain at bonding and TSV for change in (a) TSV diameter and diameter difference and (b) pitch. (Reproduced with permission from Hwang, S. et al., *Journal of Electronic Materials*, 41, 232–240, 2012.)

The increase in strain on the TSV with the rise in TSV diameter is due to the large amount of Cu at the edge of the TSV. Moreover, it can be concluded that in consideration of thermal strain generation in the bonding, the diameter difference between the TSV and the bonding plays a vital role.

Another important factor that governs the magnitude of thermal strain at the bonding is the pitch size. From Figure 2.20b, it can be observed that the strain at the bonding increases with an increase in pitch size, whereas the TSV strain is almost independent of the pitch size. Cu has a relatively large modulus of elasticity and is mainly responsible for withstanding the strain generated at the bonding. The strain can be generated from various sources such as from PCB to the 3D IC structure. The per-unit volume reduction of Cu at the bonding due to increase in pitch size increases the strain at the bonding. This is because the same amount of strain will now have to be withstood by the reduced amount of Cu in the bonding.

The effect of the TSV height and the number of die stacks on the development of thermal strain in the TSV is shown in Figure 2.21a and at the bonding is shown in Figure 2.21b. It can be observed that the effect of the TSV height and the number of die stacks on the strain developed in both TSV and bonding is insignificant. This is because the underfill (see Figure 2.19), which has relatively very small modulus of elasticity, acts as a shield against the change in strain [68].

2.5.3 Mechanical Performance

Various stresses generated due to downsizing of 3D IC may result in TSV extrusion, cracks, and so on. The TSV crack occurrence is extremely deteriorating toward the reliability of the 3D IC system and should be avoided. The understanding of mechanical properties of TSVs plays a critical role in designing crack-free TSVs under practical working conditions.

Commonly used filler materials in TSVs are Cu and W due to their higher electrical conductivity. However, these filler materials have different coefficients of thermal expansion (CTEs) from that of Si (see Table 2.1), and thus may result in generation of stress due to CTE mismatch [69]. Therefore, it is vital to study the stress induced by different TSV filler materials in order to

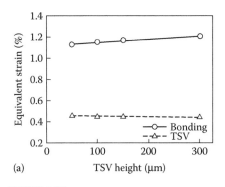
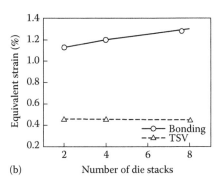

FIGURE 2.21
Change in equivalent strain at bonding and TSV as a function of (a) TSV height and (b) number of die stacks. (Reproduced with permission from Hwang, S. et al., *Journal of Electronic Materials*, 41, 232–240, 2012.)

TABLE 2.1

CTE for Different Materials

Material	CTE in 10^{-6}/K at 20°C
Si	3
Cu	17
W	4.5

Source: Dao, T. et al., *IEEE International Conference on IC Design and Technology*, 2009.

select the appropriate filler materials for providing mechanically stable conditions in 3D IC structures.

Figures 2.22 and 2.23 show an ultraviolet (UV)–Raman display of biaxial stress for a set of adjacent TSVs that have an inter-TSV spacing (S) of 10 μm and TSV lengths (L) of 30 and 10 μm, respectively. By comparing these two figures, it can be observed that the stress developed is significantly dependent

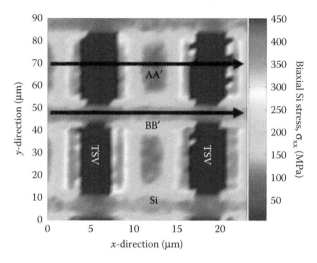

FIGURE 2.22
UV–Raman display of biaxial stress for a set of adjacent TSVs having L = 30 μm and S = 10 μm. (Reproduced with permission from Dao, T. et al., *IEEE International Conference on IC Design and Technology*, 2009.)

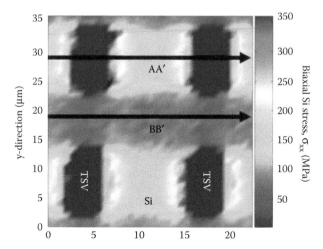

FIGURE 2.23
UV–Raman display of biaxial stress for a set of adjacent TSVs having $L = 10$ μm and $S = 10$ μm. (Reproduced with permission from Dao, T. et al., *IEEE International Conference on IC Design and Technology*, 2009.)

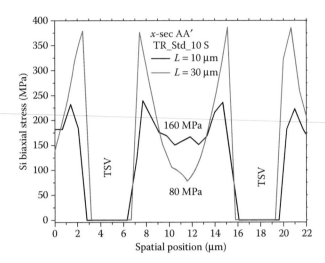

FIGURE 2.24

Cross-sectional stress along AA' direction for a set of TSVs having S = 10 μm, and L = 30 and 10 μm. (Reproduced with permission from Dao, T. et al., *IEEE International Conference on IC Design and Technology*, 2009.)

on the length of the TSVs. Figure 2.24 shows the cross-sectional stress along AA' direction for a set of TSVs of L = 30 and 10 μm and having S = 10 μm. It can be observed that compared to the TSV of L = 10 μm, the TSV of L = 30 μm experiences larger stress at the edges and shows a higher drop in the amount of stress experienced in between the TSVs. Moreover, in both cases, the drop in stress values between the TSVs seems to follow an exponential relation.

The TSV spacing, S, is now changed from 10 to 18 μm, and the stress analysis is performed for the set of TSVs of lengths 30 and 10 μm, as shown in Figure 2.25. It is observed that a similar trend as reported for S = 10 μm can be observed, in which the longer TSVs demonstrate higher stress near the edge of TSVs and sharper drop in stress between the TSVs compared to TSVs of shorter lengths. Moreover, the overstress experienced by TSVs along AA' direction decreases for higher S values. The biaxial stress along the BB' direction for the set of TSVs of lengths 30 and 10 μm, and for both cases of S = 10 and 18 μm, is shown in Figures 2.26 and 2.27, respectively. It can be observed that the stress along the BB' direction does not change as abruptly as the change in stress along the AA' direction. However, a similar relation in stress analysis can be observed as in the case of stress along the AA' direction. The TSVs of longer lengths experience higher overall stress for the same S values, and the TSVs of the same lengths experience lower overall stress for larger S values. Therefore, it can be concluded that the TSV stress depends on the TSV length and inter-TSV distance.

From this analysis, it can be observed that the region of the TSV and the substrate interface is responsible for problems relating to mechanical reliability due to maximum stress induced in that region. 3D finite-element analysis (FEA) models representing realistic TSV structures are created for carrying out stress

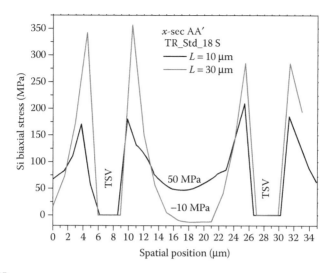

FIGURE 2.25
Cross-sectional stress along AA′ direction for a set of TSVs having $S = 18$ μm, and $L = 30$ and 10 μm. (Reproduced with permission from Dao, T. et al., *IEEE International Conference on IC Design and Technology*, 2009.)

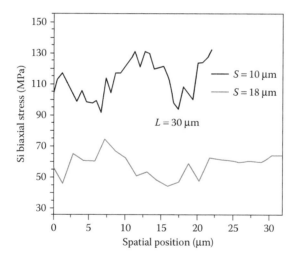

FIGURE 2.26
Biaxial stress along the BB′ direction for a set of TSVs having $L = 30$ μm, and $S = 10$ and 18 μm. (Reproduced with permission from Dao, T. et al., *IEEE International Conference on IC Design and Technology*, 2009.)

analysis near TSV/substrate–interface region. The structure of TSV is demonstrated in Figure 2.28. The 45 nm technology is considered; in Figure 2.28, 4X and 3X TSVs represent the TSV cells that occupy four and three cell rows, respectively, and KOZ signifies the keep-out zone, that is, the distance within which no cell can be placed. The liner material is considered as SiO_2.

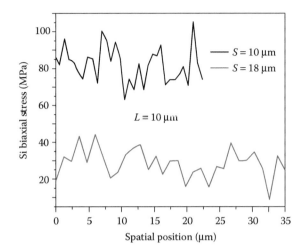

FIGURE 2.27
Biaxial stress along the BB' direction for a set of TSVs having $L = 10$ μm, and $S = 10$ and 18 μm. (Reproduced with permission from Dao, T. et al., *IEEE International Conference on IC Design and Technology*, 2009.)

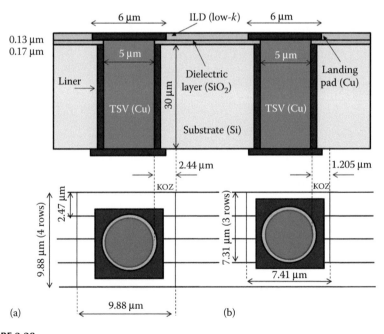

FIGURE 2.28
The baseline structure of TSV: (a) 4X TSV cell with KOZ = 2.44 μm and (b) 3X TSV with KOZ = 1.205 μm. (From Jung, M. et al., *Proceedings Design Automation Conference*, pp. 188–193, 2011. With Permission.)

The 3D FEA simulation results of normal stress induced at various regions for increase in radial distance from the center of TSV is shown in Figure 2.29. It can be observed that the 2D solution provides an incorrect stress analysis. This is because, according to the 2D solution, the stress value is highest within the TSV. However, in reality the stress value within the TSV should not be higher, because the interface region between the different materials experiences the most stress [70]. Moreover, it can be observed that both the SiO_2 liner and the landing pad decrease the stress at the TSV edge. Similar stress analysis, while considering SiO_2 and BCB as liners and for liner thicknesses (t) of 125 and 500 nm, is shown in Figure 2.30. It can be observed that with the increase in the liner thickness, the stress value at the TSV edge

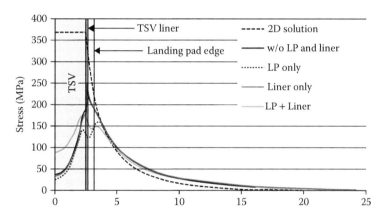

FIGURE 2.29
The effect of TSV structures on normal stress. (From Jung, M. et al., *Proceedings Design Automation Conference*, pp. 188–193, 2011. With Permission.)

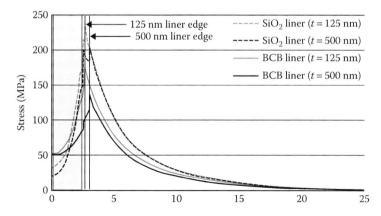

FIGURE 2.30
The effect of liner material/thickness on normal stress. (From Jung, M. et al., *Proceedings Design Automation Conference*, pp. 188–193, 2011. With Permission.)

decreases. Moreover, the BCB is a better liner material for suppressing stress compared to SiO_2. This is because Young's modulus of elasticity of BCB is much lower than that of SiO_2 [70].

2.5.4 Thermomechanical Properties

The CTE mismatch between the different materials may result in generation of stress and strain. Thermomechanical reliability is a major concern in TSVs of 3D IC. The presence of even small defects around TSVs may result in generation of stress in TSVs. This can ultimately lead to cracking of TSVs and/ or interface regions between different materials at the periphery of TSVs, as shown in Figure 2.31 [71]. The cracks generated pose great threat on the electrical, mechanical, and thermal reliabilities of the chip as a whole and may even ultimately result in malfunction of the chip [72]. Therefore, a study of thermomechanical properties of TSVs becomes necessary to avoid the unwanted reliability concerns in 3D ICs.

A schematic representation of the simplified TSV under consideration for thermal stress analysis is shown in Figure 2.32a. The details about the boundary conditions utilized are shown in Figure 2.32b. From Figure 2.32a, it should be noted that direction 1 is the radial direction, whereas direction 2 is the axial direction. Cycles of linear temperature variation is considered from −40°C to 125°C and 125°C to −40°C.

During the temperature cycling, each linear temperature increases from −40°C to 125°C, and results in approximately 5 times and 10 times more expansion of Cu compared to Si and SiO_2, respectively [73]. As a result, a deformation is produced due to temperature cycling as shown in Figure 2.33a and b. It can also be observed that the SiO_2 layer is highly strained as a result of both axial and radial expansion of Cu. However, Si is not highly strained due to higher stiffness than both Cu and SiO_2.

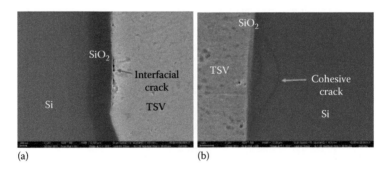

(a)   (b)

FIGURE 2.31
Growth of crack due to thermomechanical strength: (a) interfacial crack between the silicon substrate and the dielectric liner and (b) cohesive crack in the silicon substrate. (Reproduced with permission from Liu, X. et al., *Microelectronics Reliability*, 53, 70–78, 2013.)

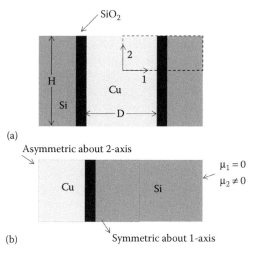

(a)

(b)

FIGURE 2.32
(a) Schematic of simplified TSV and (b) boundary condition utilized.

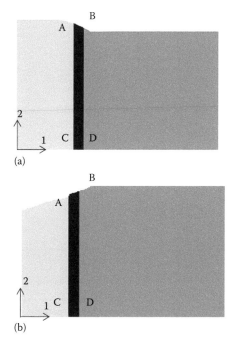

(a)

(b)

FIGURE 2.33
Exaggerated 100× deformation in TSV when (a) heating from –40°C to 125°C and (b) cooling from 125°C to –40°C.

The points A, B and points C, D are the critical points where failure may occur. This is because at points A and B, during contraction, tearing action at the interface of Cu and SiO_2 is possible, whereas at points C and D, cracking of SiO_2 and/or Cu may occur. The effect of a change in the TSV diameter, where TSVs of different aspect ratios are considered, on the radial stresses at points A (Cu) and B (SiO_2) is shown in Figure 2.34a and b, respectively. It can be observed that the stress linearly increases with the TSV diameter. Moreover, by analyzing the slope of the plots, it can be deduced that a steeper increase in stress is observed for TSVs of larger aspect ratios. By comparing the results in Figure 2.34a and b, it can be observed that the stress in Cu is always less than that in SiO_2. This is due to the fact that Young's modulus of elasticity of SiO_2 is greater than that of Cu.

The effect of change in the TSV aspect ratio on the axial strains at points (Cu) and D (SiO_2) is shown in Figure 2.35a and b, respectively. It can be observed

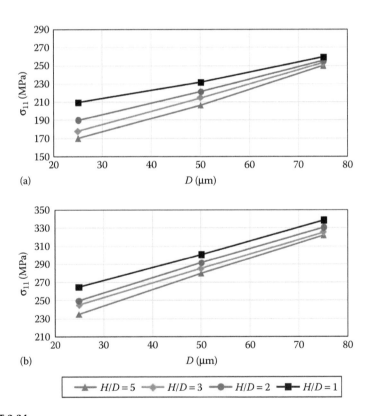

(a)

(b)

$H/D = 5$ $H/D = 3$ $H/D = 2$ $H/D = 1$

FIGURE 2.34
The effect of change in the TSV diameter on the radial stresses at (a) point A (Cu) and (b) point B (SiO_2).

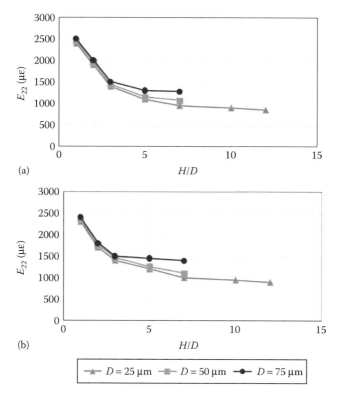

FIGURE 2.35
The effect of change in the aspect ratio on the axial stresses at (a) point C (Cu) and (b) point D (SiO_2).

that for all diameters and aspect ratios, the axial strains in Cu and SiO_2 are approximately of the same value. This is because of the close proximity of the two materials. Moreover, it can be observed that with the increase in the aspect ratio of the TSV, the axial strain decreases nonlinearly, and beyond the aspect ratio of 5, the strain becomes almost constant.

2.6 Fabrication of TSVs

The cost-effectiveness of designing a TSV follows the full manufacturing sequence that requires integration and optimization between traditional steps in back-end (BE) packaging and wafer processing. The entire flow can be optimized to deliver the greatest performance (yield, reliability) for the highest productivity. During the fabrication process, the TSV stack undergoes thinning, bonding, wafer processing on bonded/thinned wafers, and subsequent debonding. Wafers are mainly bonded to the glass or dummy silicon and thinned down to a thickness ranging from 30 to 125 μm. There are three widespread techniques used for fabrication of TSV: via-first, via-middle, and via-last.

2.6.1 Via-First TSV

In this method, TSVs are fabricated before the front-end (FE) process, that is, before the transistors are patterned in silicon as shown in Figure 2.36a [62]. The TSVs fabricated with the via-first technique experiences high temperatures during the fabrication of transistors (around 1000°C). Therefore, the TSV materials must have a lower CTE. The conventional materials such as copper (Cu; 17 ppm/K) and tungsten (W; 4.6 ppm/K) have large CTEs and must be avoided in the via-first process. Polysilicon is typically used as the filling material due to its ability to withstand higher temperatures and lesser contaminations. However, the TSV resistance of polysilicon material is large enough to limit its usage that forces the TSV fabrication designers to avoid the via-first method. Using the via-first scheme, one can choose the smallest diameter of 5–10 μm and higher aspect ratio of 10:1.

2.6.2 Via-Middle TSV

In this method, TSVs are fabricated after FE, but before BE process, that is, after the transistors are fabricated, but before the metallization layers are

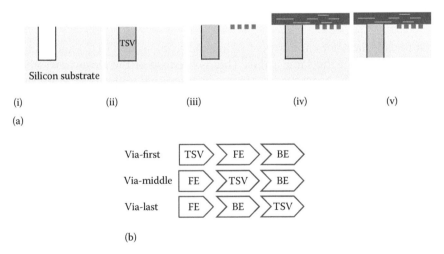

FIGURE 2.36
(a) TSV fabrication steps in the via-first method: (i) etching, insulating layer, and barrier layer deposition; (ii) TS V fill; (iii) FE process (transistors); (iv) BE process (metallization layers); and (v) wafer thinning. (b) TSV fabrication at various processing steps.

patterned as shown in Figure 2.36b. The via-middle process prefers the usage of materials such as copper or tungsten. The fabrication of on-chip metallization layers requires around 400°C in which tungsten is typically used as the filler material due to its lower CTE than Cu.

2.6.3 Via-Last TSV

In this method, TSVs are fabricated after FE and BE processes. Because the high-temperature FE and BE processes are performed before TSV formation, Cu can be used as the filler material due to its low resistivity. However, the diffusion of material in the silicon substrate and electromigration effects at high frequency are added limitations in the usage of Cu. All these issues make the system unreliable. Therefore, the current research works have been forced to choose carbon-based materials (such as CNT and graphene) with lower CTE, resistivity, electromigration effect, and contamination.

The difference between the via-first, via-middle, and via-last processes is shown in Figure 2.36b in terms of TSV placement stage. It is observed that the via-last process is the best choice for fabricating the TSVs with the traditional Cu filler material. Moreover, this approach has suitable applications for image sensors and stacked dynamic random access memory (DRAM). The primary reason behind this approach is the minimum electrical resistance of vias. For instance, a CMOS image sensor exhibits via diameters exceeding 40 µm having an aspect ratio of 2:1. In other devices, the via diameters are in the range of 10–25 µm with an aspect ratio of 5:1.

A detailed description related to different steps of TSV fabrication are discussed in Sections 2.6.4 through 2.6.9.

2.6.4 Etching

Before filling the TSV material, it is required to etch the trenches deep into the silicon that exhibits an extremely high aspect ratio. The approximate height and width of a trench are 100–150 and 1–5 µm, respectively. The etching process of TSV primarily follows a number of methods, such as wet etching, dry etching, laser drilling, reactive ion etching (RIE), and deep RIE (DRIE) [74]. Anisotropic wet etching is generally used for TSV formation with very large pitch. However, isotropic wet etching can be combined with dry etching to adjust the desired profile. Dry etching primarily includes laser drilling and RIE. Laser drilling provides significant cost advantages over patterning and etching by eliminating the lithography steps.

The RIE process usually involves a high-density plasma source in which the ions and radial species from plasma etch the surface chemically and physically. After etching, the remaining by-products are removed and the plasma becomes continuously reactive. Due to the slow nature of the RIE etching process, scientist Bosch patented a new alternative etching process known as DRIE or the Bosch process [75]. This process follows an alternative etching (using SF_6) and sidewall passivation step (using C_4F_8) with a high etch rate up to 10 μm/min. DRIE exhibits excellent process controllability and creates the vias with a high aspect ratio up to 110:1.

2.6.5 Deposition of Oxide

After completing the etching process, a trench is found in silicon. By filling the material with metal, silicon is diffused and the signal will get lost. The problem of diffusion between metal and silicon can be avoided using the oxide insulator in this trench. Accordingly, the same thickness of the oxide is required to be imposed at the top and bottom of the trench, as shown in Figure 2.37.

Plasma-enhanced chemical vapor deposition (PECVD) and atomic layer deposition (ALD) are used to deposit the oxide and nitride layers, respectively [74]. These approaches provide good adhesion to silicon at lower temperatures, around 100°C and 200°C. Oxide layers deposited through ALD have low breakdown voltage. Therefore, high-quality PECVD is desirable for oxide layer deposition in TSVs.

For dielectric applications in TSV-based interconnects, the chemical vapor deposition (CVD) is an ideal process. This process exhibits an inherently conformal deposition that is a critical coverage for subsequent titanium metal barrier and copper seed steps. To achieve high aspect ratio applications, a highly conformal deposition process is required. The deposited film produced by the CVD process exhibits good breakdown voltage and leakage current properties with excellent adhesion to the industry standard barrier metals (Ta, Ti, and TaN/Ta).

2.6.6 Barrier Layer or Seed Layer

A metal barrier layer or seed layer is placed after the deposition of the isolation layer as depicted in Figure 2.38. The purpose of these layers is to prevent the diffusion of metal into the silicon or the oxide layer. During the application of the barrier layer, it is required to deposit the oxide to obtain similar thicknesses on the top and the sides. In the process flow of TSV

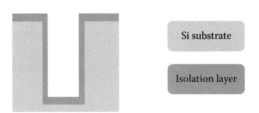

FIGURE 2.37
Deposition of isolation layer on the silicon substrate.

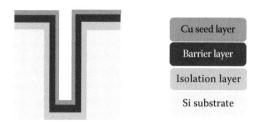

FIGURE 2.38
Growth process of Cu seed layer.

fabrication, the most challenging and expensive task is the deposition of barrier layer and subsequent filling of via.

The most favorable barrier materials are titanium and tantalum. A PVD process is used to deposit these materials that can deliver highly uniform step coverage and sheet resistance. It results in a thinner barrier for depositing a metal layer with void-free, lower stress, and good electrical conductivity [74]. Other approaches such as ALD, CVD, and electroless plating can also be applied to produce a high-quality barrier/seed layer.

2.6.7 Via Filling/Plating

Tungsten is generally used to fill the via, which is a well-known process in the wafer fabrication community. In the 1980s, tungsten was first introduced in wafer fabrication processing to fill 1 μm diameter contacts and vias up to 2 μm deep [74]. The filling requires a contact and a barrier layer that is normally sputtered titanium. The filling of tungsten is conformal, which means that more than half of the via diameter must be deposited so that the via will be filled with tungsten. It can be referred to as the standard wafer fabrication process with a relatively high throughput.

The most critical part in via fabrication is the metallization or plating step. In the via-first approach, tungsten and polysilicon are the most used conductive filler materials for TSVs. However, the conductivity of tungsten and polysilicon is lower than that of copper. However, using these materials, the plating can be made void-free under minimum stress effect. Therefore, stress is minimized during plating.

The most used process for plating is electrodeposition of copper [76]. Electrodeposition is a well-known semiconductor process that is widely used to form a conducting path. This deposition process exhibits good processability and availability at room temperature. However, it suffers from several complexities, such as reliability, throughput and process controllability [77]. Especially, high-aspect-ratio TSVs with void-free conductive TSV metal cores are difficult to implement. Therefore, an investigation is required to find some alternative approaches for plating processes. Some of these probable approaches are the use of solder balls [78], filling with conductive metal pastes [79], wire-bonded gold cores, and so on.

2.6.8 Chemical Mechanical Polishing

After metallization, chemical mechanical polishing (CMP) is done to remove the undesirable oxide or metal. This method requires a two-phase polishing approach. The first phase uses a higher removal rate of oxide with good polarization and low nonuniformity. This phase does not concern the low dishing. The second phase follows the process with a lower removal rate and acceptable dishing that is usually selective to the barrier [80]. The different steps of CMP for a Cu-based TSV are presented in Figure 2.39. This polishing process suffers from several challenges that include rapid removal of thick materials without compromising wafer topography.

2.6.9 Wafer Thinning

The thinning process of wafers generally follows two steps: (1) depositing metal films on the back to promote the backside contact to the device and (2) reducing the TSV depth. Silicon substrates can be thinned after being bonded together to allow the interconnect to be formed on the back of the bonded wafer or to reveal an existing interconnect for bonding to another substrate as shown in Figure 2.40.

Backgrinding is a generic process for thinning the wafer. In this process, a "backgrind tape" of an approximate thickness of 100–300 μm is applied to

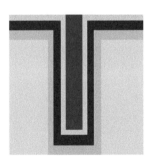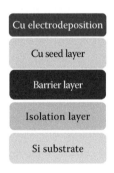

FIGURE 2.39
Cu-based TSV structure after CMP process.

FIGURE 2.40
Cu-based TSV structure after wafer thinning.

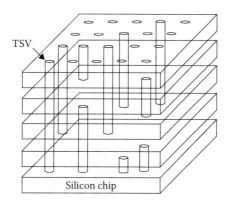

FIGURE 2.41
TSV chip stacking.

the front to protect the front side of the device wafer [80]. A grinding wheel/ disk with embedded diamonds is used to perform the grinding. The grinding is done in the following two steps: (1) a coarse grind to remove the bulk of the silicon and (2) a fine grind for thinning the wafer. For instance, a grit size of 50 μm is thinned up to 10–20 μm by using coarse grind, whereas it is reduced to 2 μm by using fine grind [80].

The TSVs manufactured by these process steps are used to connect multiple 3D dies, as shown in Figure 2.41.

2.7 Challenges for TSV Implementation

In spite of the various advantages, the technology of TSVs in 3D IC is an emerging one and is in its early stages of development. Design, test, and verification challenges are yet to be resolved, standard definitions are to be proposed, and the ecosystem of supply chain is unstable [81]. Some of the fabrication-related issues associated with TSVs may limit their applications. These issues are primarily associated with the cost, designing, testing, warpage occurrence, and manufacturing.

2.7.1 Cost

The primary barrier associated with the TSV technology is its implementation cost. The cost is mainly determined by numerous aspects of designing and manufacturing process. Particularly, the major cost barriers are the bonding/ debonding and filling materials of TSV. A TSV exhibits its tremendous value if its production cost perfectly fits with the industrial roadmap. For instance, a TSV-based 3D IC can provide improved electrical characteristics to address the limits in leakage and electrical performance of CMOS ICs beyond 11–16 nm [63]. However, these benefits come at a very high cost that can be regarded as a major challenge.

2.7.2 Design

In the current applications of 3D vertical interconnections, TSVs can be adopted to combine different types of chips, ICs, and design guidelines to address a variety of issues. The layout and chip architecture have faced fundamental design changes, such as stacking and wire bonding, due to the several thousand interconnections between the dies. Designers for each chip type used in the integration scheme will have to leverage the same master layout to line up connection points between the chips. At the same time, possible heat generation issues also need to be considered. Overheating in the stacked chips causes some designing problem related to thermal management mechanisms. Thus, the hot spots and temperature gradients strongly affect the reliability of the chip. Apart from this, longer TSV substantially increases the thermal and intrinsic stress, which becomes a major concern in mechanically stable design. The thermomechanical stress arises from the difference between the CTE of silicon and the interconnection metal, whereas the intrinsic stress results from different physical mechanisms that take place during the metal deposition. Therefore, more sophisticated solutions are required to address the designing issues related to thermal and stress management of a 3D IC [63].

From a design perspective, TSVs in 3D ICs do not require extensive retooling. There are no major bottlenecks in the process technology. A new designing of 3D system is not needed to be acquired [81]. However, new capabilities are required in areas of floorplanning and route, architectural analysis, signal integrity, thermal analysis, timing, test, and IC/package codesign.

Many TSV-based 3D ICs are capable of combining analog/radiofrequency and digital circuitry. As a result, they require strong mixed/analog signal capabilities. Due to the complex packaging requirement of the stacked die, a package/IC codesign capability is a necessity. Expertise from the field of analog, IC, digital, PCB, and package design is a must to handle TSV-based 3D ICs. Some special tools are required by the TSVs. IC design tools are required to design TSVs in active layers. Also, SiP or packaging tools may be required to plan the active layer TSVs. However, passive layer TSVs can be both designed and planned by SiP and packaging tools [81].

In 3D ICs, compared to the logic gates and other features, TSVs are very large in size. Therefore, the location and number of TSVs are critical. With the increase in the number of TSVs, the wire length also increases. TSVs very close to each other result in high crosstalk generation. Crosstalk effect can be reduced by avoiding the placement of TSVs within the keep-out zones; however, this results in additional requirement of the area. TSVs also cause generation of mechanical stress around their surrounding areas, which can have a deteriorating impact on the performance of the surrounding devices.

Routing, placement, and synthesis of 3D ICs involves a number of new challenges and considerations. Due to their large size, the TSVs have become a new layout feature. The implementation system supporting 3D ICs must consider each side of a die. This requires new database infrastructure and modeling, TSV-specific tools, and support for various stacking styles. Tools are required to support the power distribution in microbumps and TSVs. For example, the routing tool when handling the microbumps and TSVs should consider the thermal constraints to take care of hot spots. Moreover, the 3D floorplanner is required to consider X, Y, and Z directions. This facilitates to optimize the placement of TSVs and microbumps, and reduce the interconnect lengths, thus improving power and performance. For proper design convergence, the TSV and microbump assignments should consider the floorplans on the neighboring die.

2.7.3 Testing

TSV-based 3D ICs need to consider two levels of testing: the package test after the die assembly is placed into the package and the wafer test for the die. The distinction is that in TSV-based 3D ICs, the fabrication requires additional intermediate steps, such as TSV bonding. It results in more wafer testing steps prior to assembly and final packaging. The wafer testing in TSV-based 3D ICs is quite challenging. This is because (1) the available probing technology is not capable of dealing with the finer dimensions and pitch of TSV tips. Moreover, it is unable to deal with thousands of probes generally required by TSVs, because it is limited to dealing with only few hundred probes. (2) The probing technology generally leaves behind scrub marks on TSVs that can potentially hamper the bonding step. (3) Due to the finer TSV tips, very thin wafers are required for proper die stacking. However, the thin wafers are susceptible to physical damage when contacted by probing tools [81].

2.7.4 Warpage Occurrence

Generally, the distribution of TSVs used for interconnections in a wafer is nonuniform, which may result in chip warpage. Warpage means to bend/distort from the original. Typically, the TSVs are arranged in either the center or the periphery of a chip [82]. Due to different CTEs between the Cu and Si regions in a TSV, it may result in the development of compressive stress that can potentially damage the TSV and the chip as a whole. With the increase in the number of stacked chips, the warpage increases. It may also create additional tension on the microbumps, making them prone to cracking. Moreover, the warpage also hinders the proper alignment of chip components. The misalignment may result in additional stress generation in IC components.

2.7.5 Manufacturing

The cost-effectiveness of designing a TSV follows the full manufacturing sequence that requires integration and optimization between traditional steps in BE packaging and wafer processing. The entire flow can be optimized to deliver the greatest performance (yield, reliability) for the highest productivity (cost).

For TSV applications, via-first or via-last approaches primarily alter the sequence of manufacturing processes. In either approach, the TSV stack undergoes thinning, bonding, wafer processing on bonded/thinned wafers, and subsequent debonding. Wafers are mainly bonded to the glass or dummy silicon and thinned down to a thickness ranging from 30 to 125 μm [63].

2.8 Summary

The ever-increasing need for miniaturization, performance, and multifunctional microelectronic devices is motivating the microelectronic industry to move toward higher integration density. To provide a solution to these increasing demands, TSV-based 3D ICs have emerged. The graphene-based materials such as CNTs and GNRs have the potential to ultimately replace Cu as filler material for TSVs because they provide upgrades on various electrical, mechanical, and chemical aspects of Cu-based TSVs. Considering the importance of graphene-based materials, the historical development of graphene, CNTs, and GNRs is briefly discussed in this chapter. Moreover, the basic structure, types, various properties, and characteristics of both CNTs and GNRs are also discussed. Because the TSVs play a vital role in a 3D IC, the electrical, thermal, mechanical, and thermomechanical properties of TSVs are also highlighted. Furthermore, a detailed stepwise description of the fabrication of a TSV involving etching, deposition of oxide and barrier layer, via filling, CMP, and wafer thinning is also incorporated. Finally, this chapter focused on the various challenges that are faced during implementation of TSVs such as costing, designing, testing, warpage occurrence, and manufacturing.

Multiple Choice Questions

1. _____ are examples of bucky balls.
   a. Geodesic domes
   b. Hexagon
   c. Carbon nanotubes
   d. Atomic-force microscopy (AFM) and scanning tunneling micro-scope (STM)

2. In a bucky ball, a carbon atom is bound to how many adjacent carbon atoms?
   a. 1
   b. 2
   c. 3
   d. 4

3. The thermal conductivity of a standard SWCNT along its length is ____ W/mK.
   a. 3500
   b. 385
   c. 35,000
   d. 35

4. An MWCNT possesses electrical superconductivity up to a temperature of ____.
   a. 12K
   b. 12°C
   c. 100K
   d. 100°C

5. Who prepared and explained nanotubes for the first time?
   a. Sumio Tijima
   b. Richard Smalley
   c. Eric Drexler
   d. Richard Feynmann

6. The compressive strength of a nanotube ____ its tensile strength.
   a. is less than
   b. is greater than
   c. is equal to
   d. none

7. The electrical conductivity of a nanotube is ____ times that of copper.

   a. 10

   b. 100

   c. 1000

   d. 1/100

8. Carbon nanotubes, often called the strongest material, have which of the following properties?

   i. High electrical and thermal conductivity

   ii. Very high tensile strength

   iii. Higher lifetime

Select the correct answer using the codes given below:

   a. i only

   b. i and ii only

   c. i and iii only

   d. i, ii, and iii

9. The thermal stability of a nanotube is seen up to ____K in air.

   a. 100

   b. 1000

   c. 2000

   d. 3100

10. Which of the following is NOT a type of carbon nanotube structure?

   a. Armchair

   b. Chiral

   c. Zigzag

   d. Helix

11. Four carbon nanotubes are found interlocked with each other in a way that each has a concentric point. This structure is called

   a. Tetra-CNT

   b. MWCNT

   c. Quadruple CNT

   d. Poly-CNT

12. Which of the following is NOT a way to synthesize CNT?

   a. CVD

   b. Laser ablation

   c. Electric arc discharge

   d. Quantum dot synthesis

13. In 2004, a carbon nanotube was created at a world-record length of 4 cm. What kind of chemical vapor was used in order to create such a long carbon nanotube?

a. Silicon

b. Ethanol

c. Glycerol

d. Phospholipid

14. Which one of the following is the most common form of radiation used in photolithography?

a. Electronic beam radiation

b. Incandescent light

c. Infrared light

d. UV light

15. Vapor-phase epitaxy is based on which one of the following:

a. CVD

b. Diffusion

c. Ion implantation

d. PVD

Short Questions

1. What are the types of CNTs?

2. What are the differences between armchair and zigzag CNTs?

3. Give any two excellent properties of CNTs.

4. Why C-60 molecules are called as bucky balls? Explain the reason.

5. Compare the conductivities of copper, SWCNT, and MWCNT.

6. What are the advantages of MWCNT over SWCNT?

7. What are the advantages of GNRs over CNTs?

8. Explain the role of the thermal properties of TSVs in the design of 3D IC?

9. Explain the structure of MWCNT?

10. What are the steps involved in wafer thinning?

Long Questions

1. Explain the basic structure of GNR. What are the different types of GNR? Classify them as metallic or semiconducting.
2. Explain in detail the electrical properties of TSVs.
3. How does the leakage current vary in TSVs with temperature?
4. Explain the methods of fabrication of TSVs. Which is preferred and why?
5. Explain the C–V characteristics of TSV obtained at a particular temperature.
6. What are the challenges for the implementation of TSVs?
7. Explain the different methods of TSV etching.
8. What are CNTs? What are the different types of CNTs? Highlight the important properties of CNTs.
9. Why is oxide deposition needed during CNT fabrication? Explain the different methods of oxide deposition during TSV fabrication.
10. Explain the basic structure of MWCNT. What are the advantages of MWCNTs over SWCNTs.

References

1. Majumder, M. K., Kumari, A., Kaushik, B. K., and Manhas, S. K. 2014. Analysis of crosstalk delay using mixed CNT bundle based through silicon vias. In *IEEE Radio Frequency Integrated Circuits Symposium*, pp. 441–444.
2. Dreyer, D. R., Ruoff, R. S., and Bielawski, C. W. 2010. From conception to realization: An historial account of graphene and some perspectives for its future. *Angewandte Chemie International Edition* 49:9336–9344.
3. Schafhaeutl, C. 1840. Ueber die Verbindungen des Kohlenstoffes mit Silicium, Eisen und anderen Metallen, welche die verschiedenen Gallungen von Roheisen, Stahl und Schmiedeeisen bilden. *Journal für Praktische Chemie* 21:129–157.
4. Schafhaeutl, C. 1840. On the combinations of carbon with silicon and iron, and other metals, forming the different species of cast iron, steel, and malleable iron. *Philosophical Magazine* 16:570–590.
5. Brodie, B. C. 1859. On the atomic weight of graphite. *Philosophical Transactions of the Royal Society of London* 149:249–259.
6. Brodie, B. C. 1860. Researches on the atomic weight of graphite. *Quarterly Journal of the Chemical Society of London* 12:261–268.
7. Staudenmaier, L. 1898. Verfahren zur Darstellung der Graphitsäure. *Berichte der deutschen chemischen Gesellschaft* 31:1481–1487.
8. Kohlschütter, V., and Haenni, P. 1918. Zur Kenntnis des Graphitischen Kohlenstoffs und der Graphitsäure. *Zeitschrift für Anorganische und Allgemeine Chemie* 105:121–144.

9. Wallace, P. R. 1947. The band theory of graphite. *Physical Review* 71:622–634.
10. Ruess, V. G., and Vogt, F. H. 1948. Hochstlamellarer Kohlenstoff aus Graphitoxyhydroxyd. *Monatshefte fr Chemie* 78:222–242.
11. Boehm, H. P., Clauss, A., Fischer, G. O., and Hofmann, U. 1962. Dünnste kohlenstoff-folien. *Verlag der Zeitschrift für Naturforschung B* 17:150–153.
12. Morgan, E., and Somorjai, G. A. 1968. Low energy electron diffraction studies of gas adsorption on the platinum (100) single crystal surface. *Surface Science* 12:405–425.
13. May, J. W. 1969. Platinum surface LEED rings. *Surface Science* 17(1):267–270.
14. Blakely, J. M., Kim, J. S., and Potter, H. C. 1970. Segregation of Carbon to the (100) Surface of Nickel. *Journal of Applied Physics* 41(6):2693–2697.
15. Boehm, H. P., Setton, R., and Stumpp, E. 1986. Nomenclature and terminology of graphite intercalation compounds. *Carbon* 24:241–245.
16. Boehm, H. P., Setton, R., and Stumpp, E. 1994. Nomenclature and terminology of graphite intercalation compounds. *Pure and Applied Chemistry* 66:1893–1901.
17. Fitzer, E. Kochling, K.-H. Boehm, H. P., and Marsh, H. 1995. Recommended terminology for the description of carbon as a solid. *Pure and Applied Chemistry* 67:473–506.
18. McNaught, A. D., and Wilkinson, A. 1997. *Iupac in Compendium of Chemical Terminology—The Gold Book*, 2nd edition. Oxford: Blackwell Scientific.
19. Lu, X. K., Yu, M. F., Huang, H., and Ruoff, R. S. 1999. Tailoring graphite with the goal of achieving single sheets. *Nanotechnology* 10:269–272.
20. Lu, X. K., Huang, H., Nemchuk, N., and Ruoff, R. S. 1999. Patterning of highly oriented pyrolytic graphite by oxygen plasma etching. *Applied Physics Letters* 75:193–195.
21. Jang, B. Z., and Huang, W. C. 2006. Nano-scaled graphene plates. US Patent US 7071258 B1.
22. Novoselov, K. S., Geim, A. K., Morozov, S. V. et al. 2004. Electric field effect in atomically thin carbon films. *Science* 306(5696):666–669.
23. Kroto, H. W., Heath, J. R., O'Brien, S. C., Curl, R. F., and Smalley, R. E. 1985. C60: Buckminsterfullerene. *Nature* 318:162–163.
24. Monthioux, M., Serp, P., Flahaut, E. et al. 2010. Introduction to carbon nanotubes. In *Handbook of nano-technology*, ed. B. Bhushan, pp. 47–118. New York: Springer.
25. Xu, T., Wang, Z., Miao, J., Chen, X., and Tan, C. M. 2007. Aligned carbon nanotubes for through-wafer interconnects. *Applied Physics Letters* 91(4):042108-1–042108-3.
26. Tersoff, J., and Ruoff, R. S. 1994. Structural properties of a carbon-nanotube crystal. *Physical Review Letters* 73:676–679.
27. Wang, N., Tang, Z. K., Li, G. D., and Chen, J. S. 2000. Single-walled 4 Å carbon nanotube arrays. *Nature* 408:50–51.
28. Hamada, N., Sawada, S. I., and Oshiyama, A. 1992. New one-dimensional conductors, graphite microtubules. *Physical Review Letters* 68:1579–1581.
29. Li, H. J., Lu, W. G., Li, J. J., Bai, X. D., and Gu, C. Z. 2005. Multichannel ballistic transport in multiwall carbon nanotubes. *Physical Review Letters* 95(8):86601.
30. Nihei, M., Kondo, D., Kawabata, A. et al. 2005. Low-resistance multi-walled carbon nanotube vias with parallel channel conduction of inner shells. In *Proceedings of the IEEE International Interconnect Technology Conference*, pp. 234–236. Burlingame, CA: IEEE.

31. Forró, L., and Schönenberger, C. 2000. Physical properties of multi-wall nanotubes. In *Topics in Applied Physics, Carbon Nanotubes: Synthesis, Structure, Properties and Applications*, ed. M. S. Dresselhaus, G. Dresselhaus, and P. Avouris. Berlin, Germany: Springer-Verlag.

32. Close G. F., and Wong, H. S. P. 2008. Assembly and electrical characterization of multiwall carbon nanotube interconnects. *IEEE Transactions Nanotechnology* 7(5):596–600.

33. Javey, A., and Kong, J. 2009. *Carbon Nanotube Electronics*. New York: Springer.

34. Shah, T. K., Pietras, B. W., Adcock, D. J. Malecki, H. C., and Alberding, M. R. 2013. Composites comprising carbon nanotubes on fiber. US Patent US8585934 B2.

35. Dresselhaus, M., Dresselhaus, G., and Avouris, P. 2001. *Carbon Nanotubes: Synthesis, Structure, Properties and Applications*. New York: Springer.

36. Hsieh, J. Y. Lu, J. M. Huang, M. Y., and Hwang, C. C. 2006. Theoretical variations in the Young's modulus of single-walled carbon nanotubes with tube radius and temperature: A molecular dynamics study. *Nanotechnology* 17:3920–3924.

37. Forro, L., Salvetat, J. P., Bonard, J. et al. 2002. Electronic and mechanical properties of carbon nanotubes. In *Science and Application of Nanotubes*, ed. D. Tománek, and R. J. Enbody, pp. 297–320. New York: Plenum Publishers.

38. Wei, C. Srivastava, D., and Cho, K. 2002. Thermal expansion and diffusion coefficients of carbon nanotube-polymer composites. *Nano Letters* 2(6):647–650.

39. Wang, Z., and Zhao, G. L. 2013. Microwave absorption properties of carbon nanotubes-epoxy composites in a frequency range of 2–20GHz. *Open Journal of Composite Materials* 3:17–23.

40. Li, H., Yin, W. Y., Banerjee, K., and Mao, J. F. 2008. Circuit modeling and performance analysis of multi-walled carbon nanotube interconnects. *IEEE Transactions on Electron Devices* 55(6):1328–1337.

41. Naeemi, A., and Meindl, J. D. 2006. Compact physical models for multiwall carbon-nanotube interconnects. *IEEE Electron Device Letters* 27(3):338–340.

42. Naeemi, A., and Meindl, J. D. 2008. Performance modelling for single- and multiwall carbon nanotubes as signal and power interconnects in gigascale systems. *IEEE Transactions on Electron Devices* 55(10):2574–2582.

43. Fujita, M., Wakabayashi, K., Nakada, K., and Kusakabe, K. 1996. Peculiar localized state at zigzag graphite edge. *Journal of the Physics Society Japan* 65(7):1920–1923.

44. Nakada, K., Fujita, M., Dresselhaus, G., and Dresselhaus, M. S. 1996. Edge state in graphene ribbons: Nanometer size effect and edge shape dependence. *Physical Review B* 54(24):17954–17961.

45. Echtermeyer, T. J., Lemme, M. C., Baus, M., Szafranek, B. N., Geim, A. K., and Kurz, H. 2008. Nonvolatile switching in graphene field-effect devices. *IEEE Electron Device Letters* 29(8): 952–954.

46. Lemme, M. C., Echtermeyer, T. J., Baus, M., and Kurz, H. April 2007. A graphene field-effect device. *IEEE Electron Device Letters* 28(4):282–284.

47. Ouyang, Y., Yoon, Y., Fodor, J. K., and Guo, J. 2006. Comparison of performance limits for graphene nanoribbon and carbon nanotube transistors. *Applied Physics Letters* 89(20):203107-1–203107-3.

48. Naeemi, A., and Meindl, J. D. 2007. Conductance modeling for grapheme nanoribbon (GNR) interconnects. *IEEE Electron Device Letters* 28(5):428–431.

49. Li, H., Xu, C., Srivastava, N., and Banerjee, K. 2009. Carbon nanomaterials for next-generation interconnects and passives: Physics, status, and prospects. *IEEE Transactions Electron Devices* 56(9):1799–1820.

50. Kan, E. Li, Z., and Yang, J. 2011. Graphene nanoribbons: Geometric electronic and magnetic properties. In *Physics and Applications of Graphene – Theory, INTECH*, ed. S. Mikhailov, Chapter 16. Rijeka, Croatia: InTechOpen.
51. Avouris, P. 2010. Graphene: Electronic and photonic properties and devices. *Nano Letters* 10(11):4285–4294.
52. Murali, R., Brenner, K., Yang, Y., Beck, T., and Meindl, J. D. 2009. Resistivity of graphene nanoribbon interconnects. *IEEE Electron Device Letters* 30(6):611–613.
53. Dresselhaus, M. S., and Dresselhaus, G. 2002. Intercalation compounds of graphite. *Advances in Physics* 51(1):1–186.
54. Kumar, V. R., Majumder, M. K., and Kaushik, B. K. 2014. Graphene based on-chip interconnects and TSVs—Prospects and challenges. *IEEE Nanotechnology Magazine* 8(4):14–20.
55. Naeemi, A., and Meindl, J. D. 2009. Compact physics-based circuit models for graphene nanoribbon interconnects. *IEEE Transactions on Electron Devices* 56(9):1822–1833.
56. Naeemi, A., and Meindl, J. D. 2008. Electron transport modeling for junctions of zigzag and armchair graphene nanoribbons (GNRs). *IEEE Electron Device Letters* 29(5):497–499.
57. Berger, C., Song, Z. M., Li, X. et al. 2006. Electronic confinement and coherence in patterned epitaxial graphene. *Science* 312(5777):1191–1196.
58. Kumar, V. R., Majumder, M. K., Kukkam, N. R., and Kaushik, B. K. 2015. Time and frequency domain analysis of MLGNR interconnects. *IEEE Transactions on Nanotechnology* 14(3):484–492.
59. Benedict, L. X., Crespi, V. H., Louie, S. G., and Cohen, M. L. 1995. Static conductivity and superconductivity of carbon nanotubes—Relations between tubes and sheets. *Physical Review B Condens Matter* 52(20):14935–14940.
60. Xu, C., Li, H., and Banerjee, K. 2009. Modeling, analysis, and design of graphene nano-ribbon interconnects. *IEEE Transactions on Electron Devices* 56(8):1567–1578.
61. Hanlon, L. R., Falardeau, E. R., and Fischer, J. E. 1977. Metallic reflectance of AsF5-graphite intercalation compounds. *Solid State Communications* 24(5):377–381.
62. Kaushik, B. K., Majumder, M. K., and Kumar, V. R. 2014. Carbon nanotube based 3-D interconnects—A reality or a distant dream. *IEEE Circuits and Systems Magazine* 14(4):16–35.
63. Kaushik, B. K., Majumder, M. K., and Kumari, A. 2014. Fabrication and modelling of copper and carbon nanotube based through-silicon via. In *3D Circuit and System Design: Multicore Architecture, Thermal Management, and Reliability*, ed. R. Sharma, pp. 203–233. Boca Raton, FL: CRC Press, Taylor & Francis Group.
64. Katti, G., Mercha, A., Stucchi, M. et al. 2010. Temperature dependent electrical characteristics of through-Si-via (TSV) interconnections. In *Proceedings of IEEE Conference on IITC*, pp. 1–3. Burlingame, CA: IEEE.
65. Sze, S. M. 1981. *Physics of Semiconductor Devices*. New York: Wiley.
66. Huang, C., Pan, L., Liu, R. and Wang, Z. 2014. Thermal and electrical properties of BCB-liner through-silicon vias. *IEEE Transactions on Components, Packaging and Manufacturing Technology* 4(12):1936–1946.
67. Lau, J. H., and Tang, G. 2009. Thermal management of 3D IC integration with TSV (through silicon via). In *Proceedings of IEEE Electronic, Components & Technology Conference*, pp. 635–640. San Diego, CA: IEEE.

68. Hwang, S., Kim, B., Lee, H., and Joo, Y. 2012. Electrical and mechanical properties of through-silicon vias and bonding layers in stacked wafers for 3D integrated circuits. *Journal of Electronic Materials* 41(2):232–240.

69. Dao, T., Triyoso, D. H., Petras, M., and Canonico, M. 2009. Through silicon via stress characterization. In *IEEE International Conference on IC Design and Technology*.

70. Jung, M., Mitra, J., Pan, D. Z., and Lim, S. K. 2011. TSV stress-aware full-chip mechanical reliability analysis and optimization for 3D IC. In *Proceedings Design Automation Conference*, pp. 188–193. New York: IEEE.

71. Liu, X., Sundaram, V., Tummala, R. R., and Sitaraman, S. K. January 2013. Failure analysis of through-silicon vias in free-standing wafer under thermalshock test. *Microelectronics Reliability* 53(1):70–78.

72. Jung, M. G., Liu, X., Sitaraman, S. K., Pan, D. Z., and Lim, S. K. 2011. Full-chip through-silicon-via interfacial crack analysis and optimization for 3D IC. In *IEEE/ACM International Conference on Computer-Aided Design (ICCAD)*, pp. 563–570.

73. Selvanayagam, S., Lau, J. H., Zhang, X., Seah, S. K. W., Vaidyanathan, K., and Chai, T. C. 2008. Nonlinear thermal stress/strain analyses of copper filled TSV (through silicon via) and their flip-chip microbumps. In *Proceedings—Electronic Components and Technology Conference*, pp. 1073–1081. Lake Buena Vista, FL: IEEE.

74. Xu, Z., and Lu, J. Q. 2013. Through-silicon-via fabrication technologies, passives extraction, and electrical modeling for 3-D integration/packaging. *IEEE Transactions on Semiconductor Manufacturing* 26(1):23–34.

75. Laermer, F., and Schilp, A. 2003. Method of anisotropic etching of silicon. US Patent 6,531,068 B2.

76. Nilsson, P., Ljunggren, A., Thorslund, R., Hagstrom, M., and Lindskog, V. 2009. Novel through-silicon via technique for 2d/3d SiP and interposer in low-resistance applications. In *Proceedings of IEEE 59th Electronic Components and Technology Components (ECTC '09)*, pp. 1796–1801. San Diego, CA: IEEE.

77. Garrou, P., Bower, C., and Ramm, P. 2008. *Handbook of 3D Integration Technology and Application of 3D Integration Circuits*, p. 153. Weinheim, Germany: Wiley-VCH.

78. Gu, J., Pike, W. T., and Karl, W. J. 2010. A novel vertical solder pump structure for through-wafer interconnects. In *Proceedings of IEEE 23rd International Conference on Micro Electro Mechanical Systems (MEMS 2010)*, pp. 500–503. Wanchai, Hong Kong: IEEE.

79. Motoyoshi, M. 2009. Through-silicon via (TSV). *Proceedings of the IEEE* 97(1):43–48.

80. Hosali, S., Smith, G., Smith, L., Vitkavage, S., and Arkalgud, S. 2008. Through-silicon via fabrication, backgrind, and handle wafer technologies. In *Wafer Level 3-D ICs Process Technology Integrated Circuits and Systems*, eds. C. S. Tan, R. S. Gutmann, and L. R. Reif, pp. 1–32. Berlin, Germany: Springer-Verlag.

81. 3D ICs with TSVs—Design challenges and requirements, Cadence https://www.cadence.com/rl/resources/white_papers/3dic_wp.pdf.

82. Tu, K. N. 2011. Reliability challenges in 3D IC packaging technology. *Microelectronics Reliability* 51:517–523.

3

Copper-Based Through Silicon Vias

3.1 Introduction

The microelectronics industry has recently laid the foundation for the burgeoning advancement of information technology. Each electronic product launched is soon replaced by a more sophisticated successor that features enhanced functionality, connectivity, efficient power utilization, generally lower cost, and reduced size. Evolution of laptops is a good example that confirms this trend. Such enhancements have become so evident that the consumers expect the next-generation devices to be an upgrade over their previous versions in all aspects. To meet such anticipations, the electronics manufacturers are in continuous pursuit for higher integrations in order to incorporate more and more functionality within chips. Moreover, to reduce the power consumption and cost of the high-functionality chips, size reduction becomes essential.

The contribution of process technology scaling in semiconductors has played a major role in reducing the interconnection and transistor sizes over the past few decades. More than a billion transistors on a single chip have become a reality. However, such a high level of integration demands similar levels of design complexities. It is observed that the desired rate of scaling is not being achieved for transistor channel lengths of less than 20 nm [1]. This is because the process technology scaling is facing challenges such as increased leakage current and power consumption, process variability, lithography limitations, and reduced production yield. In spite of the immense efforts being made to overcome these issues, it has become evident that scaling alone cannot keep up with the required trend of integration. Development in chip packaging technologies has provided assistance toward increased integration. The development of chip packaging from its early days to the present enhancements has been discussed in Chapter 1. The advantages provided by both process technology scaling and chip packaging techniques have almost reached their maximum potential. However, increasing the demand for maintaining the trend of developments in electronic products is swiftly becoming a necessity in our lives. Therefore, it has become evident that an alternate technology to develop microelectronic systems is utmost required.

The development of three-dimensional integrated circuits (3D ICs) with through silicon vias (TSVs) has provided a promising platform to satisfy the

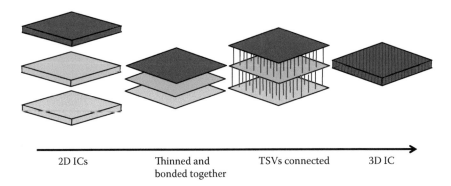

2D ICs   Thinned and   TSVs connected   3D IC
                      bonded together

FIGURE 3.1
A detailed step-by-step view showing that a 3D IC is made from 2D ICs that are reduced in thickness, bonded together, and finally connected via Cu-based TSVs.

development in accordance with Moore's law. The Cu filler material used in TSVs plays an important role in providing the desired quality of TSVs. Cu provides lower stress, void-free filling, and good thermal cycling performance, conductivity, and electrical current-carrying capability [2]. Some of the other TSV filler materials used are conductive polymer pastes, gold (Au), polysilicon, and tungsten (W) [3–5]. However, there are issues of testing, packaging, and fabrication of these filler materials. The development of a 3D IC with Cu-based TSVs from two-dimensional integrated circuits (2D ICs) is shown in Figure 3.1. TSVs provide electrical connections between oxide or through silicon tiers in vertical directions. They decrease the distance between connections and thus provide reduced signal delay and power consumption, higher bandwidth, and compact and reliable connection paths along with easing design complexities. It is important to discuss the various aspects of Cu-based TSVs because Cu is the most common filler material because of its economic feasibility and excellent electrical properties.

The rest of the chapter is divided into four sections. Section 3.2 provides the details of the physical configuration of Cu-based TSVs. Section 3.3 discusses the various equivalent electrical models of Cu-based TSVs such as TSVs with bumps, metal semiconductor (MES) ground structure, and ohmic contact in silicon interposer. Section 3.4 analyzes the performance of Cu-based TSVs in terms of propagation delay, power dissipation, crosstalk-induced delay, frequency response, and bandwidth. Finally, Section 3.5 provides a summary of this chapter.

3.2 Physical Configuration

The 3D stacking technology plays a significant role in deciding the geometry of the TSV. The different TSV physical structures such as circular, tapered, annular, square, and rectangular are shown in Figure 3.2. To evaluate the

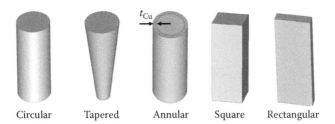

FIGURE 3.2
Different physical structures of TSVs.

shape impacts on the performance of TSVs, we must keep the cross-sectional areas constant on the top plane. Xu et al. [5] analyzed the performance evaluation between the different TSV shapes by comparing the loss parameters. The return loss and insertion loss are shown in Figure 3.3a and b, respectively. It can be observed among all the shapes of TSVs, the tapered shape TSV has the smallest return loss and insertion loss. This is due to the fact that the tapered shape TSV has a relatively low capacitance value because of its reduced surface area. To reduce the fabrication cost, the annular structured TSVs are preferable due to less amount of filler material needed for their fabrication.

The TSVs can be used for power supply/ground, clock, or analog and digital signals [1]. In general, the geometry of all TSVs within a chip is the same to provide ease in optimizing the 3D technology process. After choosing a TSV technology, the only flexibility in the electrical characteristics of TSV design is the use of more TSVs in parallel to each other within the same link. This can be helpful in power supply networks that require low-resistance interconnects and high DC current capability. Connecting the TSVs in parallel results in the decrease of resistance in power/ground links of the stacked dies. It also results in the increase of the TSV parasitic capacitance. The TSV structure, in terms of both materials and geometry, significantly contribute to the parasitic values of the TSV.

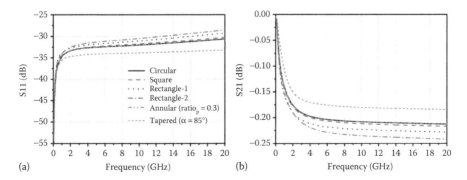

FIGURE 3.3
Impact of TSV shapes on electrical performance (a) S11 and (b) S21. (From Xu, Z. and Lu, J.-Q., *IEEE Transactions on Components, Packaging and Manufacturing*, 1, 154–162, 2011. With Permission.)

The placement of TSVs within 3D ICs is a complex process of additional process steps and may affect the reliability of the chip. The typical range of TSV height is several hundreds of micrometers and that of TSV diameter is several tens of micrometers. Discussing the physical configuration of TSVs along with involving physical characterization will bring out a clear understanding of TSV structures. The fabrication of TSVs first involves the creation of holes into the Si by etching. The liner oxide (SiO_2) is then deposited into the hole followed by the diffusion barrier layers (TaN/Ta or TiN/Ti). The important Cu seed layers are then deposited. To properly fill the TSVs with Cu, electrochemical deposition is used. For Cu recrystallization, thermal annealing is performed. This step is required to provide resistance against electromigration and stress-generated degradation, and to improve the conductance of the TSVs. To isolate the TSVs and remove the excess Cu, chemical–mechanical polishing is performed. After processing, it is required that the surface of the wafer in the region around the TSVs should be flat in order to properly integrate the TSVs within the chips. Moreover, it is essential to assess the reliability of the diffusion barriers within TSVs. This is because the metal contamination in transistors, particularly in the active regions, can result in failure or severe performance degradations of electrical devices.

The uniformity in cross section of TSVs is an important criterion for providing overall reliability. As the TSVs are quite large in size, their preparation can be very time consuming, requiring several hours for every cross section. The etch profile can be evaluated by a cleave through the TSV [6]. A good-quality cross-sectional preparation is difficult if the metal fill in TSVs is completed. This is because the metal layers will be ripped out or deformed during cleaving. The cross section is required for such samples for the purpose of measuring the diffusion barrier layer thickness. The cross sections in such cases can be prepared via the techniques of broad-beam argon ion milling, focused ion beam milling, or mechanical polishing [6].

3.3 Modeling of Cu-Based TSVs

3D field solvers require high computational time; therefore, it is essential to develop an electrical equivalent model of TSV to reduce the computational time and for a better physical insight. Moreover, the equivalent models have to be analytical, upgradable, reducible, and must consider the physical dimensions and material properties; hence, if the material properties or the dimensions are scaled, the model gets adapted to the changed parasitic values accordingly. This allows for easier implementation and modifications in the TSV design structures.

3.3.1 Scalable Electrical Equivalent Model of Coupled TSVs with Bumps

A scalable electrical equivalent model of coupled TSVs is presented in this section. As the model is obtained through the physical structure, each parasitic component is clearly expressed through its physical meaning with the help of the closed-form equations. To obtain the scalable model, the analytical equations are derived based on the material properties and structural parameters.

Kim et al. [7] proposed a high-frequency scaled electrical model for the analysis of coupled TSVs. The proposed model considers not only the TSVs but also the bumps that are the additional components for 3D-based TSV design. The structure of signal and ground TSVs with bumps are shown in Figure 3.4, in which the bumps provide connection between the stacked heterogeneous dies. Depending on the configuration, the electrical equivalent model of a pair of Cu TSVs is shown in Figure 3.5 [7].

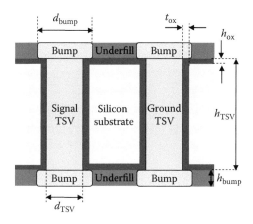

FIGURE 3.4
Structure of signal TSV and ground TSV with bumps.

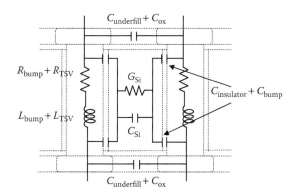

FIGURE 3.5
Electrical equivalent model of signal and ground TSVs.

The resistances of the TSV and bump are represented by R_{TSV} and R_{bump}, respectively. At high frequencies, the current flows primarily on the surface of the conductor with an exponential reduction in current density toward the depth of the conductor. Therefore, the overall resistance of the conductor is substantially reduced at high frequencies. The depth of penetration in which the current falls off to e^{-1} of its value near the surface is known as skin depth (δ). This skin depth is accounted for in the modeling of R_{TSV} and R_{bump} at high frequencies. The resistances are determined by the resistivity of the TSV and bump, ρ_{TSV} and ρ_{bump}; the radius of the TSV and bump, r_{TSV} and r_{bump}; and the height of the TSV and bump, h_{TSV} and h_{bump}, respectively. The skin depth determines the material properties such as permittivity, resistivity of the TSV filler material, and the operating frequencies. The proximity factor, k_p, is determined by the ratio of distance between the two TSVs (p_{TSV}) and their radii (r_{TSV}) [8]. The resistance of the TSV can be expressed as

$$R_{TSV} = \sqrt{\left(R_{dc,TSV}\right)^2 + \left(R_{ac,TSV}\right)^2} \tag{3.1}$$

where:

$$R_{dc,TSV} = \rho_{TSV} \times \frac{h_{TSV}}{\pi \times r_{TSV}^2}$$

$$R_{ac,TSV} = k_p \left(\rho_{TSV} \times \frac{h_{TSV}}{2\pi \times r_{TSV} \times \delta_{TSV} - \pi \delta^2_{TSV}} \right) \tag{3.2}$$

$$\delta_{TSV} = \sqrt{\frac{\rho_{TSV}}{\pi \times f \times \mu_{TSV}}} \tag{3.3}$$

The resistance of the bump is expressed as

$$R_{bump} = \sqrt{\left(R_{dc,bump}\right)^2 + \left(R_{ac,bump}\right)^2} \tag{3.4}$$

where:

$$R_{dc,bump} = \rho_{bump} \times \frac{h_{bump}}{\pi \times r_{bump}^2}$$

$$R_{ac,bump} = k_p \left(\rho_{bump} \times \frac{h_{bump}}{2\pi \times r_{bump} \times \delta_{bump} - \pi \delta^2_{bump}} \right) \tag{3.5}$$

$$\delta_{bump} = \sqrt{\frac{\rho_{bump}}{\pi \times f \times \mu_{bump}}} \tag{3.6}$$

The parasitic inductance has started to play an important role in the analysis of TSV performance due to the adoption of low-resistance TSV materials and high operating switching frequencies. The TSV inductance (L_{TSV}) and bump inductance (L_{bump}) are expressed as [9]

$$L_{TSV} = \frac{1}{2}\left[\frac{\mu_0\mu_{r,TSV}}{2\pi} \times h_{TSV} \times \ln\left(\frac{p_{TSV}}{r_{TSV}}\right)\right] \qquad (3.7)$$

$$L_{bump} = \frac{1}{2}\left\{\frac{\mu_0\mu_{r,bump}}{2\pi} \times h_{bump} \times \ln\left(\frac{p_{bump}}{r_{bump}}\right)\right\} \qquad (3.8)$$

To avoid the signal flow from the TSV conductor to the silicon substrate, an insulating layer surrounding the TSV metal layer is necessary. Thus, an insulating capacitance is formed between the metal TSV and the semiconductor substrate. The $C_{insulator}$ is determined by the h_{TSV}, the height of the insulating layer h_{ox}, r_{TSV}, and the permittivity of the insulating layer. Using the coaxial cable capacitance model, the $C_{insulator}$ is derived as

$$C_{insulator} = \frac{1}{2}\left[2\pi \times \varepsilon_0\varepsilon_{r,insulator} \times \frac{h_{TSV} - 2h_{ox}}{\ln\left(\frac{r_{TSV} + t_{ox}}{r_{TSV}}\right)}\right] \qquad (3.9)$$

Due to the overlapping region of bump-to-silicon substrate of the upper side as well as the lower side, a capacitance is formed named as C_{bump}, and it must be added to the $C_{insulator}$ as shown in Figure 3.5. The C_{bump} is modeled as a parallel plate capacitor and can be expressed as

$$C_{bump} = \varepsilon_0\varepsilon_{r,insulator} \times \frac{\pi \times \left[r_{bump}^2 - (r_{TSV} + t_{ox})^2\right]}{h_{ox}} \qquad (3.10)$$

The capacitance between the signal TSV and the ground TSV forms the insulating layer, C_{ox}, and the capacitance between the signal bump and the ground bump forms the underfill, $C_{underfill}$. These capacitances are modeled as parallel-wire capacitances and can be expressed as [7]

$$C_{ox} = \frac{\pi \times \varepsilon_0\varepsilon_{r,insulator}}{\cosh^{-1}\left(\frac{p_{TSV}}{2r_{TSV}}\right)} \times h_{ox} \qquad (3.11)$$

$$C_{underfill} = \frac{\pi \times \varepsilon_0\varepsilon_{r,underfill}}{\cosh^{-1}\left(\frac{p_{TSV}}{2r_{bump}}\right)} \times h_{bump} \qquad (3.12)$$

Due to the semiconductor silicon substrate, there are conductance and capacitance between the signal and ground TSVs. Using the parallel-wire capacitance model, the C_{si} is represented as

$$C_{si} = \frac{\pi \times \varepsilon_0 \varepsilon_{r,si}}{\cosh^{-1}\left(\dfrac{p_{TSV}}{2r_{TSV}}\right)} \times \left(h_{TSV} - 2h_{ox}\right) \quad (3.13)$$

Using the relationship between the capacitance and the conductance, $C_{si}/G_{si} = \varepsilon_{si}/\sigma_{si}$ [10], the conductance can be expressed as

$$G_{si} = \frac{\pi \times \sigma_{si}}{\cosh^{-1}\left(\dfrac{p_{TSV}}{2r_{TSV}}\right)} \times \left(h_{TSV} - 2h_{ox}\right) \quad (3.14)$$

Because all the parasitic elements are expressed in terms of physical dimensions and material properties, the model can be used to estimate the TSV behavior with different material properties and dimensions.

3.3.2 Modeling of Multicoupled TSVs

In a realistic scenario, we come across multiple numbers of TSVs; consider a TSV that is present in the vicinity of the reference TSV pair as shown in Figure 3.6. The conventional formulas derived in Section 3.3.1 cannot be applied for this configuration because the substrate capacitance varies in presence of other TSVs. Therefore, this section is focused on the modeling of substrate capacitance and conductance of multicoupled TSVs. C. R. Paul [11] considered the similar problem and calculated the capacitance matrix for ribbon cables. Using the same approach, the capacitance and conductance matrix can be calculated as

$$[C] = \mu_0 \varepsilon_{Si} [L]^{-1} \quad (3.15)$$

$$[G] = \mu_0 \sigma_{Si} [L]^{-1} \quad (3.16)$$

where L is the inductance matrix and each element is expressed as

$$L_{ii} = \frac{\mu_0}{2\pi} \ln\left(\frac{d_{i0}^2}{r_i r_0}\right) \quad (3.17)$$

$$L_{ij} = \frac{\mu_0}{2\pi} \ln\left(\frac{d_{i0} d_{j0}}{d_{ij} r_0}\right) \quad (3.18)$$

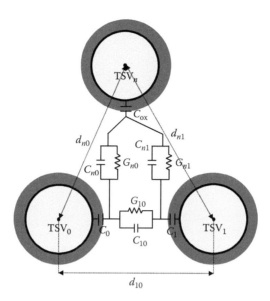

FIGURE 3.6
Conductance and capacitance modeling of coupled multiple TSVs.

3.3.3 Modeling of Coupled TSVs with MES Ground Structure

The conventional TSV structure discussed in Section 3.3.1 is the metal–insulator–semiconductor (MIS) configuration. Process development on TSVs are now gearing up toward the new type of configuration named as MES type as shown in Figure 3.7, and the top view of the MES structure is shown in Figure 3.8. In the MES configuration, the insulating layer around the ground TSV is removed as shown in Figure 3.7. The MES ground TSV behaves as a ground plug to reduce the noise in the substrate [12]. Additionally, these MES TSVs behave like a thermal via that can effectively transfer heat. However, the MES TSVs are not preferred for signal and power TSVs, because they deteriorate due to the formation of an ohmic contact to the silicon substrate. Engin et al. [13] observed that the MES TSVs completely eliminate the capacitive crosstalk in a low-frequency mode. However, the reduction in crosstalk is less significant for shorter TSV heights, because the line resistance of the ground TSV dominates over coupling parasitics.

3.3.4 Modeling of TSVs with Ohmic Contact in Silicon Interposer

Engin et al. [13] assumed a perfect ohmic contact between the metal and the silicon substrate as discussed in Section 3.3.3. In general, the silicon substrate of an IC is usually grounded, and therefore, the contact resistance and capacitance at the interface of metal–silicon can be safely ignored. However, because the silicon interposer is not often grounded, the resistances ($R_{contact}$ and R_{region}) and capacitance (C_{doping}) need to be

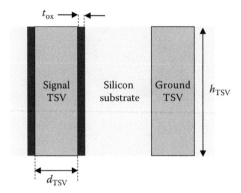

FIGURE 3.7
Cross-sectional view of MIS signal TSV and MES ground TSV.

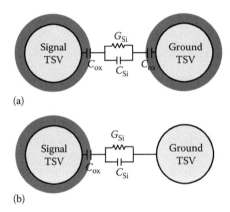

FIGURE 3.8
Top view of (a) conventional MIS structure and (b) modified structure with MIS signal TSV and MES ground TVS.

included during the modeling of TSVs in silicon interposer. This section describes the effect of contact resistance, doping resistance, and capacitance on the performance of MES TSVs [14]. A signal–ground–signal (SGS) TSV is shown in Figure 3.9. The ground TSV is used as a shield between the two signal TSVs by creating a Faraday cage [15]. However, the ground TSV with doping region around it provides a better signal reference to the signal TSVs due to significantly reduced contact resistance. Therefore, during the fabrication of TSVs, a doping region is created around the ground TSV to reduce its contact resistance. Thus, the doping resistance (R_{region}) and capacitance (C_{doping}) must be incorporated into the electric equivalent model as shown in Figure 3.10.

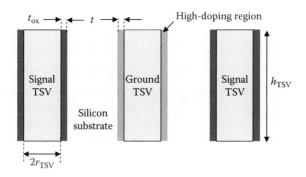

FIGURE 3.9
Signal–ground–signal TSV structure with doping region around the ground TSV.

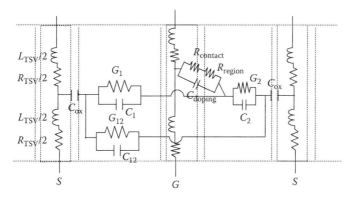

FIGURE 3.10
Electrical equivalent model of signal–ground–signal TSV with ohmic contact around the ground TSV.

The contact resistance must be added to the R_{region} due to the work function difference between the metal and the semiconductor and can be modeled as

$$R_{\text{contact}} = \frac{\rho_c(N_a)}{2\pi r_{\text{TSV}} h_{\text{TSV}}} \tag{3.19}$$

where:
h_{TSV} is the height of the TSV
$\rho_c(N_a)$ is the contact resistivity that is obtained from the doping concentration

The resistance and capacitance of the doping region can be expressed as

$$R_{\text{region}} = \frac{\rho}{2\pi h_{\text{TSV}}} \ln\left(\frac{r_{\text{TSV}} + t}{r_{\text{TSV}}}\right) \tag{3.20}$$

$$C_{\text{doping}} = \frac{\varepsilon_{\text{Si}} h_{\text{TSV}}}{\ln\left(\dfrac{r_{\text{TSV}} + t}{r_{\text{TSV}}}\right)} \qquad (3.21)$$

The performance of SGS TSVs with and without doping regions was analyzed in [15]. It is observed that the doping region around the ground TSV significantly reduces the near-end coupling noise. This reduction is primarily dependent on the concentration profile of the doping region.

3.4 Performance Analysis of Cu-Based TSVs

This section analyzes the performance of Cu-based TSVs for different via heights. The performance of Cu-based TSVs is compared in terms of propagation delay, power dissipation, crosstalk-induced delay, frequency, and bandwidth analysis. Circuit simulations are performed based on the electrical equivalent model to investigate the performance analysis of TSVs.

3.4.1 Propagation Delay and Power Dissipation

The schematic view of Cu-based TSV that is driven by a complementary metal oxide semiconductor (CMOS) driver and terminated by a capacitive load is shown in Figure 3.11. The material properties considered are shown in Table 3.1. The equivalent load capacitance is considered as 45 aF. The propagation delay and power dissipation for varying heights of Cu-based TSV are shown in Figure 3.12.

It can be observed from Figure 3.12 that as the TSV height increases, the delay and power dissipation also increase. There is almost a linear dependence of delay and power dissipation with the TSV height. This is because the TSV parasitics increase with the TSV height, and both the delay and the power dissipation linearly depend on the TSV parasitics.

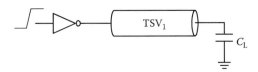

FIGURE 3.11
Schematic view of Cu-based TSV driven by a CMOS driver and terminated by capacitive load.

TABLE 3.1

Material Properties of Cu-Based TSVs

Symbol	Value	Symbol	Value
σ_{Si}	$10\,S/m$	$\mu_{r,RDL}$	1
$\varepsilon_{r,ox}$	4	$\mu_{r,Bump}$	1
$\varepsilon_{r,Si}$	11.9	$\mu_{r,TSV}$	1
ρ_{RDL}	$1.68 \times 10^{-8}\,\Omega m$	$\varepsilon_{r,Underfill}$	7
ρ_{Bump}	$1.68 \times 10^{-8}\,\Omega m$	$\varepsilon_{r,ox,bot}$	4
ρ_{TSV}	$1.68 \times 10^{-8}\,\Omega m$	$\varepsilon_{r,IMD}$	4

Source: Kim, J. et al., *IEEE Transactions on Components, Packaging and Manufacturing Technology*, 1(2), 181–195, 2011.

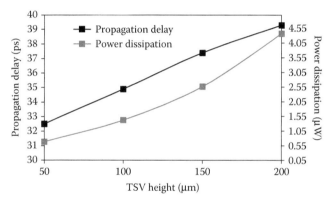

FIGURE 3.12
Propagation delay and power dissipation of Cu-based TSVs for varying heights.

3.4.2 Crosstalk-Induced Delay

The capacitive coupling is generated between the two TSVs, if they are placed in close proximity to each other [16]. Figure 3.13 shows two Cu-based TSVs in close proximity to each other. As a result, the capacitive coupling is generated as shown in Figure 3.14. The crosstalk noise is analyzed in the following two cases: (1) in-phase delay, in which the coupled TSVs are simultaneously switched in the same phase and (2) out-phase delay, in which the coupled TSVs are simultaneously switched in the opposite phase.

The crosstalk-induced propagation delay is analyzed at different TSV heights and shown in Figure 3.15. The mutual capacitance is considered in the line capacitance during out-phase switching of coupled TSVs, and it is neglected during the in-phase switching of coupled TSVs. This is due to the effect of Miller coupling capacitance. The Miller capacitance largely

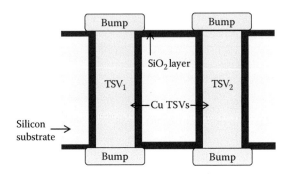

FIGURE 3.13
Physical configuration of a pair of Cu-based TSVs.

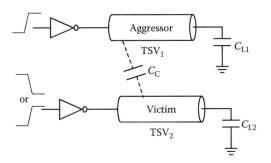

FIGURE 3.14
Schematic view of capacitively coupled TSV lines.

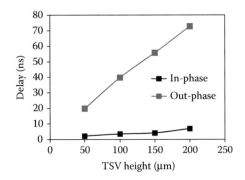

FIGURE 3.15
Comparison of crosstalk-induced dynamic in-phase and out-phase propagation delay between coupled TSVs.

influences the signal propagation when the coupled lines transit in the opposite direction, whereas it has no effect when the coupled lines transit in the same direction [17]. Therefore, it can be observed that the out-phase delay is higher than the in-phase delay.

The nonlinear dependence of propagation delay on the TSV height can be observed from Figure 3.15. The delay time of the driver–TSV–load (DTL) system is dependent on the parasitics of the driver, the TSV line, and the load. The TSV parasitics are linearly dependent on the height of the TSV, whereas the driver and load parasitics are not dependent on the TSV height. Also, the driver response to different TSV lines and load parasitics is nonlinear. Therefore, the cumulative effect of the TSV height on the delay time is nonlinear [17].

3.4.3 Frequency Response and Bandwidth Analysis

This section analyzes the frequency response and bandwidth of Cu-based TSVs with the help of transfer function (TF) of the DTL system, as shown in Figure 3.16. The TF accurately considers the driver resistance, the driver capacitance, and the via parasitic, as shown in Table 3.2.

To obtain the overall gain (V_{out}/V_{in}), the DTL of Figure 3.16 is represented as a cascaded connection of several two-port networks, as shown in Figure 3.17. The best choice of the two port parameters is the *ABCD* parameter. The *ABCD* matrix parameters of each two-port network are represented as $g_1, g_2,$

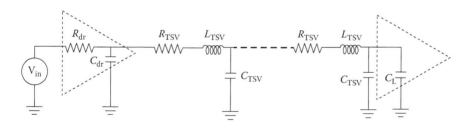

FIGURE 3.16

A DTL system. R_{dr} and C_{dr} represent the driver resistance and capacitance, respectively.

TABLE 3.2

Equivalent Parasitics of Cu-Bundled TSVs

Via Parasitics	Values
R_{TSV} (mΩ/μm)	7.1
L_{TSV} (pH/μm)	1.05
C_{TSV} (fF/μm)	0.45
R_{dr} (Ω)	292.36
C_{dr} (aF)	45.51

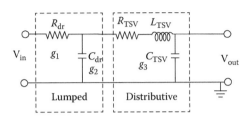

FIGURE 3.17
Cascaded connection of Figure 3.16.

and g_3. Telegrapher's equation of a distributed transmission line is used to obtain the *ABCD* matrix parameter (gain) g_2.

Using the transmission matrix parameter for a uniform RLC transmission TSV of height h, the resultant matrix parameter of the DTL configuration can be expressed as [18,19]

$$T_{result} = g_1 \bullet g_2 \bullet g_3$$

$$= \begin{bmatrix} 1 & R_{dr} \\ 0 & 1 \end{bmatrix} \begin{bmatrix} 1 & 0 \\ sC_{dr} & 1 \end{bmatrix} \begin{bmatrix} \cosh(\gamma nx) & Z_0 \sinh(\gamma nx) \\ 1/Z_0 \sinh(\gamma nx) & \cosh(\gamma nx) \end{bmatrix} \quad (3.22)$$

$$= \begin{bmatrix} A & B \\ C & D \end{bmatrix}$$

In the above equation, Z_0, γ, n, and x are the characteristic impedance, propagation constant, number of distributed segments, and length of each segment, respectively.

Now, using Equation 3.22, the distributed TF of the DTL can be expressed as [17]

$$\frac{V_{out}}{V_{in}} = TF = \frac{1}{A + sC_L B} \quad (3.23)$$

where:
  $C_L = 10$ fF is the load capacitance
  A and B are the coefficients

The derivation of the coefficients A and B can be derived from Equation 3.22.

The TF of Equation 3.23 is used to obtain the cutoff frequency (f_c) that primarily depends on the via parasitics. The parasitic values are obtained for a fixed via radius of 2.5 μm. The second-order TF of the DTL system can be expressed as

$$TF = \frac{1}{a_0 + a_1 s + a_2 s^2} \tag{3.24}$$

where the coefficients a_0, a_1, and a_2 can be derived from Equation 3.23. Substituting $a_0 = 1$, the roots of Equation 3.24 can be expressed as

$$s_1 = \frac{-a_1 + \sqrt{a_1^2 - 4a_2}}{2a_2} \tag{3.25}$$

and

$$s_2 = \frac{-a_1 - \sqrt{a_1^2 - 4a_2}}{2a_2} \tag{3.26}$$

Therefore, the cutoff frequency, f_c, can be obtained as

$$f_c = \frac{1}{2\pi} \sqrt{\frac{(2a_2 - a_1^2) + \sqrt{(2a_2 - a_1^2)^2 + 4a_2^2}}{2a_2^2}} \tag{3.27}$$

For a fixed TSV radius of 2.5 μm, Figure 3.18 shows the plots for the frequency response of DTL at TSV heights of 30, 60, 90, and 120 μm, respectively. It is

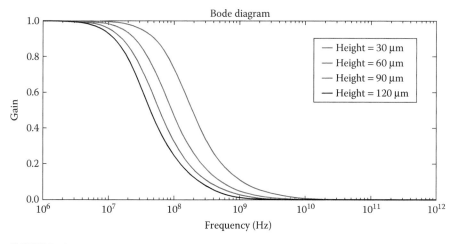

FIGURE 3.18
Frequency response of Cu-based TSVs for different TSV heights.

TABLE 3.3

Cutoff Frequency of Cu-Based TSVs for Different Heights

TSV Height (μm)	Cutoff Frequency (GHz) for Cu-Based TSVs
30	0.856
60	0.158
90	0.036.8
120	0.00912

observed that the longer TSVs exhibit smaller bandwidth. It is due to comparatively higher values of R and C with the increase in the TSV height. It leads to form an RC low-pass filter that has a bandwidth close to the cutoff frequency, $f_c = 1/2\pi RC$. Table 3.3 presents the cutoff frequency of Cu-based TSVs for different TSV heights.

3.5 Summary

TSV-based 3D IC technology has become inevitable to maintain the trend of chip development in accordance with Moore's law. The TSVs provide shorter interconnections between the chip and the substrate, thus improving the electrical performance of the system through decrease in power dissipation and propagation delay. An insight into the physical configuration along with physical characterization of Cu-based TSVs has been provided in this chapter. The typical range of the TSV height is several hundreds of micrometers and the TSV diameter is several tens of micrometers. For performance analysis, electrical modeling of Cu-based TSVs is discussed. The scalable electrical equivalent model of coupled TSVs is presented, which can be represented by the lumped model-based analytical equations. The modeling of TSVs with bumps, coupled TSVs with MES ground structure, TSVs with ohmic contact in silicon interposer is also provided. Based on the electrical modeling of TSVs with bumps, the performance analysis of the Cu-based TSVs is carried out. The performance analysis comprised the propagation delay, power dissipation, comprehensive crosstalk analysis in terms of functional, dynamic in-phase and out-phase crosstalk, and finally the frequency and bandwidth analyses. It was observed that the delay and power dissipation increase, whereas the bandwidth decreases with the TSV height.

Multiple Choice Questions

1. Which of the following material is used as a TSV filler material?
   a. Polymer paste
   b. Silicon dioxide
   c. Polysilicon
   d. All of the above

2. In which direction TSVs provide electrical connection between oxide and tiers?
   a. Horizontal direction
   b. Vertical direction
   c. No connection
   d. All directions

3. Which of the following TSV shapes has smallest insertion loss and return loss?
   a. Circular
   b. Tapered
   c. Annular
   d. Square

4. Which of the following TSV shape has the lowest fabrication cost?
   a. Rectangular
   b. Tapered
   c. Annular
   d. Circular

5. Arrange the following in the correct order of fabrication steps used in TSV fabrication:
   a. Etching =>SiO$_2$ deposition =>diffusion =>Cu seed layer deposition
   b. Cu seed layer deposition =>SiO$_2$ deposition =>diffusion =>etching
   c. Etching =>Cu seed layer deposition =>diffusion =>SiO$_2$ deposition
   d. Diffusion =>SiO$_2$ deposition =>etching =>Cu seed layer deposition

6. The process step used in preparing the cross section of TSV is
   a. Broad-beam argon ion milling
   b. Focused ion beam milling
   c. Mechanical polishing
   d. None of the above

7. The doping region is created around the ground TSV
    a. To reduce ground resistance
    b. To increase ground resistance
    c. To provide better signal reference
    d. Both b and c
8. If the height of the TSV is increased, the contact resistance would
    a. Increase
    b. Decrease
    c. Remain constant
    d. None of the above
9. If the height of the TSV is increased, the bandwidth would
    a. Increase
    b. Decrease
    c. Remain constant
    d. None of the above
10. The MES ground TSV behaves as a ground plug to reduce
    a. Noise
    b. Resistivity
    c. Conductivity
    d. Capacitance

Short Questions

1. Why the desired rate of scaling is not being achieved for transistor channel lengths of less than 20 nm?
2. Why Cu is used as a filler material in TSVs than W?
3. Why the tapered shape TSV has the smallest return loss and insertion loss?
4. What are the process steps used for filling the Cu and to recrystalline it?
5. Why the good-quality cross-sectional preparation is difficult if the metal fill in TSVs is completed?
6. Draw the power delay product curve with respect to the TSV diameter.
7. Explain how the crosstalk noise is affected with respect to pitch distance.

8. Write the expression of the cutoff frequency for DTL and explain how the bandwidth is changing with the height of TSV?

9. What is skin depth and how the resistance is affected by it?

10. What is proximity factor?

Long Questions

1. Explain the physical configuration of TSV and explain its impact on electrical performance?

2. Explain the electrical model of coupled TSV and also describe its resistance and insulated capacitance?

3. Derive the expression for substrate conductance in coupled TSVs?

4. Explain multicoupled TSV with its conductance and capacitance modeling?

5. Explain the frequency response for Cu-based TSV and derive the expression for its bandwidth.

References

1. Papanikolaou, A., Soudris, D., and Radojcic, R. 2010. *Three Dimensional System Integration: IC Stacking Process and Design*. New York: Springer.

2. Gupta, A., Kannan, S., Kim, B., Mohammed, F., and Ahn, B. 2010. Development of novel carbon nanotube TSV technology. In *IEEE Proceedings of ECTC*, June, pp. 1699–1702. Las Vegas, NV.

3. Kikuchi, H, Yamada, Y., Mossad, A., Liang, J., Fukushima, T., Tanaka, T., and Koyanagi, M. 2008. Tungsten through-silicon via technology for three dimensional LSIs. *Journal of Applied Physics* 47(4):2801–2806.

4. Xu, C., Li, H., Suaya, R., and Banerjee, K. 2009. Compact AC modeling and analysis of Cu, W, and CNT based through-silicon vias (TSVs) in 3-D ICs. In: *Proceedings IEEE International Electron Device Meeting*, pp. 521–524. Baltimore, MD: IEEE Press.

5. Xu, Z., and Lu, J.-Q. 2011. High-speed design and broadband modeling of through-strata-vias (TSVs) in 3D integration, CPMT. *IEEE Transactions on Components, Packaging and Manufacturing* 1(2):154–162.

6. Mayer, A., Grimm, G., Hecker, M., Weisheit, M., and Langer, E. 2013. Challenges for physical failure analysis of 3D-integrated devices- sample preparation and analysis to support process development of TSVs. In *Conference Proceedings from the 39th International Symposium for Testing and Failure Analysis*, November 3–7, pp. 433–436. San Jose, CA.

7. Kim, J., Pak, J. S., Cho, J. et al. 2011. High frequency scalable electrical model and analysis of a through silicon via (TSV). *IEEE Transactions on Components, Packaging and Manufacturing Technology* 1(2):181–195.
8. Terman, F. E. 1943. *Radio Engineers' Handbook*. New York: McGraw-Hill.
9. Sadiku, M. N. O. 2001. *Elements of Electromagnetics*. Oxford: Oxford University Press.
10. Hall, S. H., Hall, G. W., and McCall, J. A. 2000. *High-Speed Digital System Design: A Handbook of Interconnect Theory and Design Practices*. New York: Citeseer
11. Paul, C. R. 1978. Prediction of crosstalk in ribbon cables: Comparison of model predictions and experimental results. *IEEE Transactions on Electromagnetic Compatibility* 20(3):394–406.
12. Khan, N. H., Alam, S. M., and Hassoun, S. 2011. Mitigating TSV-induced substrate noise in 3-D ICs using GND plugs. In *Quality Electronic Design, 12th International Symposium*, March 14–16, pp. 1–6. Santa Clara, CA.
13. Engin, A. E., and Narasimhan, S. R. 2012. Modeling of crosstalk in through silicon vias. *IEEE Transactions on Electromagnetic Compatibility* 55(1):149–158.
14. Yang, D. C., Xie, J., Swaminathan, M., Wei, X. C., and Li, E. P. 2013. A rigorous model for through-silicon vias with ohmic contact in silicon interposer. *IEEE Microwave and Wireless Components Letters* 23(8):385–387.
15. Xie, B., Swaminathan, M., Han, K. J., and Xie, J. 2011. Coupling analysis of through-silicon via (TSV) arrays in silicon interposers for 3D systems. In *Electromagnetic Compatibility, 2011 IEEE International Symposium*, August 14–19, pp. 16–21. Long Beach, CA.
16. Kahng, A. B., Muddu, S., and Vidhani, D. 1999. Noise and delay uncertainty studies for coupled RC interconnects. In *Proceedings IEEE International ASIC/SOC Conference*, Washington, DC, pp. 3–8.
17. Kumar, V. R., Majumder, M. K., Alam, A., Kukkam, N. R., and Kaushik, B. K. 2015. Stability and delay analysis of multi-layered GNR and multi-walled CNT interconnects. *Journal of Computational Electronics* 14(2):611–618.
18. Fathi, D., Forouzandeh, B., Mohajerzadeh, S., and Sarvari, R. 2009. Accurate analysis of carbon nanotube interconnects using transmission line model. *IET Micro & Nano Letters* 4(2):116–121.
19. Kumar, V. R., Majumder, M. K., Kukkam, N. R. and Kaushik, B. K. 2015. Time and frequency domain analysis of MLGNR interconnects. *IEEE Transactions on Nanotechnology* 14(3):484–492.

4

Modeling and Performance Analysis of CNT-Based Through Silicon Vias

4.1 Introduction

During the recent past, several researchers [1] have designed the stacked integrated circuit (IC) layers on top of each other in order to integrate more devices on a single chip with improved performance. This advocated technique, known as three-dimensional (3D) die stacking, primarily results in higher transistor density, improved speed, lower power dissipation and area [2]. Traditionally, the connections were made through the multiple intellectual property cores on a single die (system on chip), multiple dies in a single package (multichip package), and multiple ICs on a printed circuit board [3]. Later on, system-in-package technology was introduced in which dies containing ICs are stacked vertically on a substrate. Another stacking technique is package on package that uses vertically stacked multiple packaged chips [4]. The latest development in this area is the 3D stacked IC using through silicon vias (TSVs), which employs a single package containing a vertical stack of naked dies and allows the die to be vertically interconnected with another die. TSVs are primarily referred to as a vertical electrical connection, or vertical interconnect access (via), that passes completely through a silicon wafer or a die. A TSV-based 3D IC offers various advantages in integrating a heterogeneous system onto a single platform as shown in Figure 4.1.

The performance of 3D IC is primarily dependent on the choice of filler materials used in TSVs. Tungsten (W) and copper (Cu) are the most commonly used filler materials in 3D TSVs. However, in recent years, the W and Cu have faced certain challenges due to the fabrication limitations in achieving proper physical vapor deposition, seed layer deposition, and performance limitations due to electromigration and higher resistivity. Therefore, researchers are forced to find an alternative replacement of W and Cu materials. Carbon nanotubes (CNTs) are considered to be the promising material in the current nanoscale regime due to their unique electrical, thermal, and mechanical properties [5–7]. The superior electrical properties are due to the unique band structure of graphene that leads to zero effective mass of electrons and holes. CNTs have the

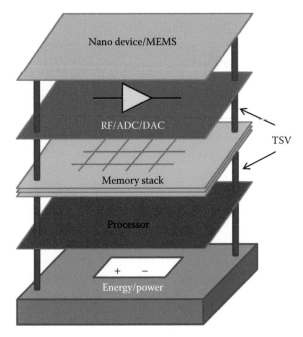

FIGURE 4.1
TSV-based 3D technology. ADC, analog to digital converter; MEMS, micro-electro-mechanical systems; and DAC, digital to analog converter.

higher current-carrying capability, long ballistic transport length, higher thermal conductivity, and better mechanical strength than Cu [8–10].

The performance of Cu- and single-walled CNT (SWCNT) bundled TSVs were previously analyzed and compared by Gupta et al. [11] for different via dimensions. The authors proposed an equivalent electrical model by neglecting the imperfect metal–nanotube contact resistance. Xu et al. [12] proposed a compact *resistance inductance capacitive conductance (RLCG)* model of SWCNT and multiwalled CNT (MWCNT) bundled TSVs for a fixed number of conducting channels and mean free paths (mfps). However, Zhao et al. [13] considered the diameter-dependent conducting channels while modeling via parasitics. The authors compared the performance of Cu-, W-, and SWCNT bundled TSVs for a perfect contact resistance ($R_{mc} = 0$) and a fixed value of mfp ($\lambda_{mfp} = 1\ \mu m$). These analyses [11–13] lacked in accuracy and did not analyze the performance thoroughly. Therefore, a more realistic model and analysis is required that can fairly compare the performance of Cu-, SWCNT bundle-, and MWCNT bundled TSVs.

This chapter analyzes and compares the propagation delay, power dissipation, crosstalk, and bandwidth of Cu-based TSVs and bundled TSVs having SWCNTs and MWCNTs of different number of shells. A comprehensive and accurate equivalent single conductor (ESC) model is employed, taking into account the metal–oxide–semiconductor (MOS) effect generated by the presence of TSVs on the silicon substrate. The via parasitics are accurately modeled by considering the diameter-dependent mfp and conducting channels. The ESC model of Cu, SWCNT bundle, and MWCNT bundle are used to represent the TSV in a driver–TSV–load (DTL) system. However, a capacitively coupled TSVs is used to analyze the crosstalk-induced delay for in-phase and out-of-phase switching scenarios.

This chapter is organized as follows: Section 4.1 introduces the recent research scenario and describes briefly about the works carried out. Section 4.2 provides a detailed description of the physical configuration and equivalent electrical models of a pair of CNT bundled TSVs. Using the equivalent electrical model of different CNT bundled TSVs, Section 4.3 analyzes the power, delay, crosstalk, and bandwidth for different TSV heights. Finally, Section 4.4 draws a brief summary of the chapter.

4.2 Physical Configuration

The stacking of TSVs on the Si substrate primarily uses an SWCNT/MWCNT bundle as a prospective filler material. A thin layer of silicon dioxide (SiO_2) is the most commonly used dielectric material for TSVs. However, for high-frequency applications, SiO_2 cannot be used due to its large fringing capacitance [14]. Hence, this layer needs to be replaced with a suitable polymer liner. The capacitance of TSVs with polymer liners reduces due to its lower dielectric constants and larger thickness than SiO_2 liners. It results in an improved electrical performance in terms of higher speed, reduced power dissipation, and crosstalk [15,16].

The physical configuration of a pair of CNT bundled TSVs is shown in Figure 4.2. The number of CNTs (N_{CNT}) is approximated using the cross-sectional area of the via and the diameter of each CNT in a bundle. For a given via radius $r_{TSV} = 45$ nm, the cross-sectional area of the TSV is equal to 6361.73 nm^2. Therefore, the total numbers of SWCNTs, double-walled

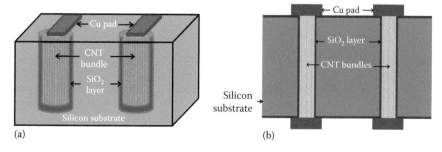

FIGURE 4.2
Schematic and (a) physical configuration of a pair of CNT bundled TSVs and (b) cross-sectional view.

CNTs (DWCNTs), and 4-, 8-, and 10-shell MWCNTs in a bundle are obtained as 8100, 2869, 876, 244, and 159, respectively, for a fixed SWCNT and DWCNT/MWCNT inner shell diameter (d_{CNT}) of 1 nm.

4.3 Real Possibilities of CNT-Based TSVs

Although SWCNT and MWCNT bundled TSVs demonstrate improved electrical, mechanical, and thermal behavior over Cu-based TSV, but they still face certain challenges to fabricate in terms of large imperfect metal–nanotube contact resistance, poor control of chirality and orientation, higher growth temperature during fabrication, and so on. This section discusses the primary challenges faced by CNT-based TSVs.

4.3.1 Imperfect Metal–Nanotube Contact Resistance

The metal–CNT contact may cause large contact resistance that is dependent on the fabrication process. During fabrication, the metal–electrode contact with CNTs may cause a reflection effect that occurs due to the inefficient coupling of the electron wave function from the electrode into the contact. The weak intertube and intershell coupling between SWCNTs in a bundle and inside MWCNTs, respectively, is the key challenge for the direct connections between all graphene shells. Therefore, forming a low contact resistance is one of the primary challenges in the fabrication of CNT-based vias.

4.3.2 Densely Packed CNT Bundles

A densely packed CNT bundled TSVs can outperform copper vias in terms of conductivity. However, it is not trivial to grow a densely packed SWCNT/MWCNT bundle with a fixed intertube spacing of 0.34 nm. The material and the size of catalyst particles are the key challenges for determining the diameter and density of nanotubes.

4.3.3 Chirality Control

The growth of CNTs inside a bundled TSV cannot control the chirality. Statistically, only one-third of SWCNTs with random chirality are metallic. An improved metallic-to-semiconducting tube ratio can only increase the conductivity of SWCNT bundled TSVs.

4.3.4 Defect-Free CNTs

CNTs are very sensitive to adsorbed molecules. It is observed that the adsorbed molecules on the surface of CNTs affect the electrical resistance, which is one of the additional technical challenges for producing CNTs with acceptable defect-free characteristics.

4.3.5 Higher Growth Temperature of CNTs

The higher growth temperature required during fabrication is another major limitation in the employment of CNT-based TSVs. Temperature more than 600°C is required during CNT growth, which is unfortunately incompatible with complementary MOS (CMOS) devices and other temperature-sensitive materials.

Challenges such as purification, separation of CNTs, control over the chirality and alignment, high-density growth, and high-quality contacts are yet to be resolved. Based on the aforementioned discussion, it can be inferred that several process- and reliability-related challenges are to be met out before CNT employment turns out to be reality.

4.4 Modeling

Development of a reliable 3D integrated system is largely dependent on the choice of filler materials used in TSVs. Several researchers [6–10] have preferred SWCNT and MWCNT bundles over Cu as prospective filler materials due to their higher conductivity. This section presents different electrical equivalent circuit models of SWCNT and MWCNT bundled TSVs.

4.4.1 Compact AC Model of SWCNT Bundled TSVs

The filler material of embedded TSV on the Si substrate is composed of different types of CNT bundles. A top view of TSV is shown in Figure 4.3 in which the SWCNT bundle (TSV filler material) is surrounded by a dielectric layer for dc isolation. The isolation dielectric is further surrounded by a depletion region. Previously, researchers [17–19] ignored the depletion region around the TSV dielectrics (MOS effect). The thickness of this depletion region primarily depends on the applied voltage bias, interface charge density, and material properties, and has a major influence on the capacitance model [20,21]. However, these authors [20,21] neglected the interface charge while modeling the depletion region. Later, Xu et al. [12] introduced a comprehensive and accurate compact $RLCG$ circuit model of a pair of SWCNT bundled TSVs as shown in Figure 4.4. This model is valid from low- to high-frequency regimes in consideration of the MOS effect on Si, skin effect in TSV, and eddy current in Si substrate.

The equivalent resistance and inductance of SWCNT bundled TSV are represented as r_{TSV} and l_{TSV}, respectively. The capacitance per unit height of

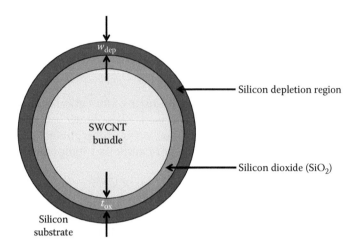

FIGURE 4.3
Top view of SWCNT bundled TSV.

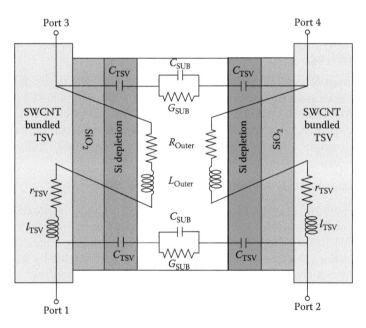

FIGURE 4.4
Equivalent *RLCG* circuit model of a pair of SWCNT bundled TSVs.

TSV, c_{TSV}, arises due to the potential difference between the SiO$_2$ and the Si depletion region [12]. The c_{TSV} can be expressed as

$$c_{TSV} = \left[\frac{1}{2\pi\varepsilon_{ox}} \bullet \ln\left(1 + \frac{t_{ox}}{r_{via}}\right) + \frac{1}{2\pi\varepsilon_{Si}} \bullet \ln\left(1 + \frac{r_{dep}}{r_{via} + t_{ox}}\right) \right]$$ (4.1)

For the dielectric layer (SiO$_2$) and the Si depletion region, ε_{ox} and ε_{Si} are the permittivities, whereas t_{ox}, r_{via}, and r_{dep} represent the oxide thickness, TSV radius, and radius of the depletion region, respectively. The resistance and inductance per unit TSV height are represented as R_{outer} and L_{outer}, respectively. The R_{outer} and L_{outer} mainly arise due to the eddy currents on the Si substrate. Therefore, both R_{outer} and L_{outer} primarily depend on the distance between two TSVs (d_{pitch}) and can be expressed as

$$R_{outer} = \frac{\omega\mu}{2} \bullet \text{Re}\left[H_0^{(2)}\left(\frac{1-j}{\delta_{Si}}\left(r_{via} + t_{ox} + W_{dep}\right) \right) - H_0^{(2)}\left(\frac{(1-j)d_{pitch}}{\delta_{Si}} \right) \right] \quad (4.2)$$

$$L_{outer} = \frac{\mu}{\pi}\ln\left(\frac{r_{via} + t_{ox} + W_{dep}}{r_{via}} \right)$$
$$+ \frac{\mu}{2} \bullet \text{Im}\left[H_0^{(2)}\left(\frac{1-j}{\delta_{Si}}\left(r_{via} + t_{ox} + W_{dep}\right) \right) - H_0^{(2)}\left(\frac{(1-j)d_{pitch}}{\delta_{Si}} \right) \right] \quad (4.3)$$

where:
μ represents the permeability of the Si substrate
$H_0^{(2)}$ and δ_{Si} are the zeroth-order Hankel function and the skin depth of the Si substrate, respectively

The effective conductance and capacitance of the Si substrate are represented by G_{SUB} and C_{SUB}, respectively, and can be expressed as

$$G_{SUB} = \frac{\pi\sigma_{Si}}{\ln\left(\frac{d_{pitch}}{2r_{via}} + \left[\sqrt{\left(\frac{d_{pitch}}{2r_{via}} \right)^2 - 1} \right] \right)} \quad (4.4)$$

$$C_{SUB} = \frac{\varepsilon_0\varepsilon_r A}{d_{pitch}} \quad (4.5)$$

where:

$$A = \pi r_{via} h_{TSV} \text{ and}$$

where σ_{Si} and h_{TSV} are the conductivity of the Si substrate and the height of TSV, respectively.

The modeling of via parasitics of SWCNT bundle primarily depends on the number of conducting channels of each SWCNT in the bundle. The total number of SWCNTs in a bundle can be obtained using the radius of single (r_{CNT}) and bundled (r_{via}) SWCNTs and the van der Waal's distance ($\delta \approx 0.34$ nm) between the two SWCNTs as shown in Figure 4.5. Thus, N_{CNT} can be expressed as

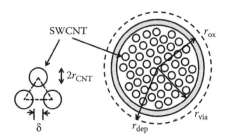

FIGURE 4.5
Cross section of an SWCNT bundle.

$$N_{CNT} = \frac{2\pi r_{via}^2}{\sqrt{3}\left(2r_{CNT}+\delta\right)^2} \tag{4.6}$$

For metallic SWCNTs in a bundle, the average number of conducting channels for a particular diameter of SWCNT can be expressed as [22]

$$
\begin{aligned}
N_i(D_i) &\approx k_1 T D_i + k_2, & D_i > d_T/T \\
&\approx 2/3 & D_i \le d_T/T
\end{aligned} \tag{4.7}
$$

where:
  D_i represents the diameter of the *i*th SWCNT in bundle
  constants k_1 and k_2 posses a value of 2.04×10^{-4} nm^{-1}K^{-1} and 0.425, respectively

The thermal energy of electrons and the gap between the two sub-bands determine the quantitative value of d_T that is equivalent to 1300 nm · K at room temperature ($T = 300$K) [22]. For $D_i > 4.3$ nm, the average number of conducting channels is proportional to its CNT diameter. Thus, the total number of conducting channels in a bundle can be expressed as the summation of conducting channels (N_i) of each SWCNT.

$$N_{channel} = \sum_{i=1}^{n} N_i \tag{4.8}$$

where *n* represents the number of CNTs in a bundle.
  The conduction mechanism in CNT is ballistic (lossless) or dissipative due to the large value of mfp in the range of micrometers. The mfp is proportional to the diameter of an individual SWCNT as [23]

$$\lambda_{mfp,i} = \frac{1000\,D_i}{\left(\dfrac{T}{T_i}\right)-2} \tag{4.9}$$

Thus, the total number of conducting channels in an SWCNT bundle can be expressed as

$$N_{\text{Total}} = N_{\text{channel}} \times N_{\text{CNT}} \tag{4.10}$$

Each SWCNT in the bundle primarily demonstrates three different types of resistances: (1) quantum or intrinsic resistance (R_q) that arises due to confinement of electrons in a nanowire, (2) imperfect metal–nanotube contact resistance (R_{mc}) that exhibits a value of few hundreds of kilo-ohms depending on the fabrication process, and (3) scattering resistance (r_{TSV}) that arises due to the higher nanotube length exceeding mfp's of electrons. Therefore, the intrinsic and per-unit height (p.u.h.) scattering resistance of the equivalent model can be expressed as [10]

$$r_{\text{TSV}} = \frac{\left[\sum_{i=1}^{n} \left(\frac{R_q}{2N_i \lambda_{\text{mfp},i}} \right)^{-1} \right]^{-1}}{N_{\text{Total}}} \tag{4.11}$$

where the quantum resistance R_q is expressed as

$$R_q = \frac{h}{2e^2} \approx 12.9\,k\Omega \tag{4.12}$$

where h and e represent Planck's constant and the charge of an electron, respectively.

The equivalent *RLC* model of CNT bundle primarily comprises two different types of capacitances: (1) the quantum capacitance (c_q^{Bundle}) that represents the finite density of states at Fermi energy and (2) the electrostatic capacitance (c_e^{Bundle}) that occurs due to the potential difference between the bundle and the ground plane. Thus, the p.u.h. c_q^{Bundle} and c_e^{Bundle} can be expressed as [24]

$$c_q^{\text{Bundle}} = c_{q0} \times N_{\text{Total}} \tag{4.13}$$

where:

$$c_{q0} = \frac{2e^2}{hv_F}$$

$$c_e^{\text{Bundle}} = \frac{2\pi\varepsilon_0}{\cosh^{-1}\left(\dfrac{d_{\text{CNT}}^{\text{Outer}} + h_{\text{TSV}}}{d_{\text{CNT}}^{\text{Outer}}} \right)} \tag{4.14}$$

where v_F and $d_{\text{CNT}}^{\text{Outer}}$ represent the Fermi velocity ($\approx 8 \times 10^5$ m/s) and the average diameter of SWCNTs facing the ground plane, respectively.

The equivalent inductance of the via (l_{TSV}) consists of (1) the kinetic inductance (l_k) that originates from the kinetic energy of electrons and (2) the magnetic inductance (l_m) that arises due to the magnetic field induced by the current flowing through a nanotube. Therefore, the p.u.h. l_{TSV}, l_k, and l_m can be expressed as [10]

$$l_{TSV} = \frac{l_k}{N_{Total}} + l_m \qquad (4.15)$$

where:

$$l_k = \frac{h}{2e^2 v_F} \text{ and } l_m = \frac{\mu}{2\pi} \ln\left(\frac{y}{d_{CNT}^{Outer}}\right) \qquad (4.16)$$

where y is the distance between the CNT bundle and the ground plane.

4.4.2 Simplified Transmission Line Model of a TSV Pair

Depending on the physical configuration of a TSV pair, Kannan et al. [24] proposed an equivalent transmission line model of SWCNT bundled TSVs as shown in Figure 4.6. The authors designed the TSVs in a coplanar waveguide fashion for measuring power at the frequency range of 2–20 GHz. Each TSV is electrically modeled by via-self inductance, resistance, and capacitance. The SWCNT bundles are used as TSV filler materials; therefore, the r_{TSV} and l_{TSV} primarily represent the equivalent resistive and inductive parasitics of the SWCNT bundle, respectively. R_{mc} is the imperfect metal–nanotube contact resistance that mainly depends on the fabrication process. The capacitance and conductance of the Si substrate, C_{SUB} and G_{SUB}, respectively, depend mainly on the center-to-center distance between two TSVs (d_{pitch}). In the equivalent circuit model, C_{TSV_OX} and C_{ox} are the capacitance of SiO_2 around TSV and the fringing capacitance between two TSVs, respectively [25]. The p.u.h. quantum and electrostatic capacitances of SWCNT bundled TSVs are represented as c_q^{Bundle} and c_e^{Bundle}, respectively.

C_{TSV_OX} and C_{OX} in Figure 4.6 primarily represent the capacitance of SiO_2 around TSV and the fringing capacitance between two TSVs, respectively, and can be expressed as

$$C_{TSV\_OX} = \frac{4\varepsilon_0 \varepsilon_r t_{Si} \left(r_{via} - t_{ox}\right)}{t_{ox}} \qquad (4.17)$$

$$C_{OX} = \left[\left(\frac{2}{C_{TSV\_OX}} + \left(\frac{\varepsilon_0 \varepsilon_r A}{d_{pitch}}\right)^{-1}\right)^{-1}\right]^{-1} \qquad (4.18)$$

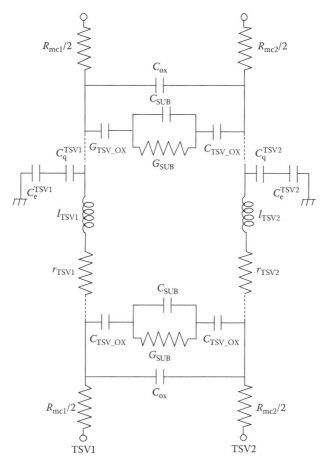

FIGURE 4.6
Equivalent electrical model of a pair of SWCNT bundled TSVs.

Kannan et al. [24] neglected the substrate-dependent resistance and inductance that adversely affect the modeling of TSVs in the high-frequency regime. Taking this into account, Gupta et al. [11] considered the substrate as a large parallel plate capacitor and modeled the associated parasitics as effective series resistance (R_{EFF}) and inductance (L_{EFF}) as shown in Figure 4.7. The model is compatible for RF application at 60 GHz frequencies.

Before moving toward the descriptive modeling, it is required to understand the parasitics that effectively characterize the SWCNT and the substrate behavior. R_{EFF} and L_{EFF} are the effective series resistance and inductance, respectively, as shown in Figure 4.7. R_{EFF} primarily represents the resistance of TSVs and inner conductors, and is analyzed using a resonant line technique. Thus, R_{EFF} of the Si substrate can be obtained as

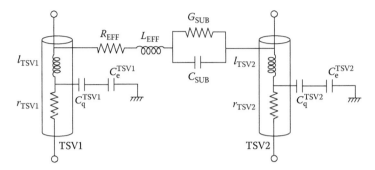

FIGURE 4.7
Equivalent electrical model of SWCNT bundled TSVs considering the substrate as a parallel plate capacitor.

$$R_{EFF} = Z_0 \left[\frac{\pi}{4} \frac{f}{f_0} \cos ec^2 \left(\frac{\pi}{2} \frac{f}{f_0} \right) - \frac{1}{2} \cot \left(\frac{\pi}{2} \frac{f}{f_0} \right) \right] \left(\frac{1}{Q_S} - \frac{1}{Q_0 \sqrt{f/f_0}} \right) + \frac{\omega L}{Q_S} - R_0 \quad (4.19)$$

where Q_S and Q_0 represent the Q-factor at natural (f) and test frequency (f_0), respectively. The resistive loss (R_0) in the Si substrate primarily depends on the frequency components and can be expressed as

$$R_0 = 0.004 \left(\frac{f}{f_0} \right)^{0.84} \quad (4.20)$$

L_{EFF} is referred as a function of TSV pitch which is inversely proportional to each other and can be expressed as

$$L_{EFF} = l_{TSV} \bullet K_g \quad (4.21)$$

where l_{TSV} and K_g are the self-inductance and the correction factor, respectively. K_g primarily depends on the diameter of TSV pitch and can be expressed as

$$K_g = K_{g\_a} - K_{g\_b} \bullet \ln \left(\frac{r_{via}}{t_{silicon} + t_{insulator}} \right) \quad (4.22)$$

where $t_{silicon}$ and $t_{insulator}$ represent the thicknesses of the Si substrate and the insulating layer (SiO$_2$), respectively. The coefficients K_{g_a} and K_{g_b} can be determined using circuit optimization during the model extraction process for the silicon substrate and the insulating layer, respectively.

G_{SUB} and C_{SUB} represent the effective conductance and capacitance of the Si substrate and can be expressed as Equations 4.4 and 4.5, respectively. r_{TSV} and l_{TSV} represent the parasitic resistance and inductance of SWCNT bundle in the equivalent model of Figure 4.7. Via parasitics such as c_q^{TSV} and c_e^{TSV} that primarily depend on the number of SWCNTs in bundle are also accounted for. Higher bundle density can be obtained with higher number of SWCNTs in a bundle. For higher number of SWCNTs, one can obtain more number of conducting channels in parallel that effectively reduce the resistance and inductance to provide a better transmission and throughput.

The equivalent electrical model proposed by Gupta et al. [26] incorrectly considered all SWCNTs in a TSV bundle as metallic. In the current fabrication, it is reported that only one-third of SWCNTs in bundle is metallic [22]. In the recent years, Zhao et al. [13] introduced an equivalent electrical model of a pair of metallic SWCNT bundled TSVs as shown in Figure 4.8a.

The simplified transmission line model of Figure 4.8a is shown in Figure 4.8b. In order to derive the simplified model, it is imperative to consider a "complex effective conductivity" that depends on the intrinsic self-impedance. The effective impedance of the pair of CNT-based TSVs can be determined as [13]

$$Z_{eff} = Zh_{TSV} = (R + j\omega L)h_{TSV} \tag{4.23}$$

The Z_{eff} depends on both the skin depths of TSV and the silicon substrate, δ_{TSV} and δ_{Si}, respectively, and can be expressed as [13]

$$\delta_{TSV} = \sqrt{\frac{2}{\omega\mu\sigma_{eff}}} \tag{4.24}$$

$$\delta_{Si} = \sqrt{\frac{2}{\omega\mu(\sigma_{Si} + j\omega\varepsilon_{Si})}} \tag{4.25}$$

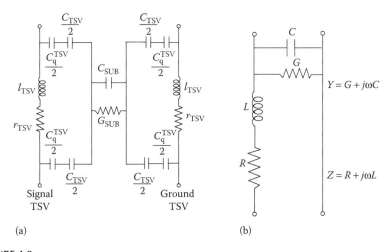

(a)   (b)

FIGURE 4.8
(a) Equivalent *RLC* model and (b) its simplified transmission line model for a pair of SWCNT bundled TSVs.

where σ_{Si}, ε_{Si}, and μ represent the conductivity, permittivity, and permeability of silicon substrate, respectively.

The elements G_{SUB} and C_{SUB} in Figure 4.8a represent the conductance and capacitance of the silicon substrate, respectively, and can be expressed as [13]

$$C_{SUB} = \frac{\pi \varepsilon_{Si} h_{TSV}}{\cosh^{-1}\left(\dfrac{d_{pitch}}{2r_{dep}}\right)} \tag{4.26}$$

$$G_{SUB} = \frac{\sigma_{Si} C_{SUB}}{\varepsilon_{Si}} \tag{4.27}$$

The effective admittance of the pair of CNT-based TSVs in Figure 4.8b can be expressed as

$$Y_{eff} = Y h_{TSV} = (G + j\omega C) h_{TSV}$$

$$= \left[2\left(j\omega c_q^{TSV}\right)^{-1} + 2\left(j\omega c_{TSV}\right)^{-1} + \left(G_{SUB} + j\omega C_{SUB}\right)^{-1} \right]^{-1} \tag{4.28}$$

where c_q^{TSV} represents the equivalent quantum capacitance of the SWCNT bundle. c_{TSV} in Figure 4.8a is the MOS capacitance in the inversion region and can be expressed as

$$c_{TSV} = \left(\frac{1}{C_{ox}} + \frac{1}{C_{dep}}\right)^{-1} = \frac{\ln\left(\dfrac{r_{ox}}{r_{TSV}}\right)}{2\pi\varepsilon_0 h_{TSV}} + \frac{\ln\left(\dfrac{r_{dep}}{r_{ox}}\right)}{2\pi\varepsilon_{Si} h_{TSV}} \tag{4.29}$$

where r_{dep} and r_{ox} are the depletion and oxide radii, respectively, as shown in Figure 4.5. Physically, the TSV pair with higher electrical conductivity has lesser conductive loss and better transmission characteristics. The electrical conductivity of SWCNT bundle is dependent on the frequency and the geometrical parameters.

4.4.3 Modeling of MWCNT-Based TSV

Sarto and Tamburrano [27] proposed a multiconductor transmission line (MTL) model of vertical MWCNT between two horizontal planes. Figure 4.9a and b presents the schematic of a vertical MWCNT between two metallic planes and the cross section of an MWCNT, respectively. The total number of shells in MWCNT primarily depends on the intershell distance ($\delta = 0.34$ nm) and the diameters of inner and outer shells. The *i*th shell of MWCNT is characterized by the number of conducting channels (N_i)

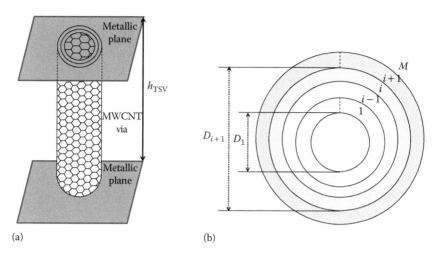

(a)   (b)

FIGURE 4.9
(a) Physical configuration of an MWCNT via and (b) cross section of an MWCNT.

of one-dimensional energy sub-bands crossing the Fermi level. The N_i is dependent on the diameter of ith shell, D_i. The MTL model of an MWCNT via is presented in Figure 4.10.

In the equivalent electrical circuit model, Z'_{via} and Y'_{via} are the matrices of external impedance and admittance of the MWCNT via, respectively. The scattering resistance, kinetic inductance, and quantum capacitance of the ith shell are represented as $r_s(i,i)$, $l_k(i,i)$, and $c_q(i,i)$, respectively. The shell voltage and current at any point x along the transmission line can be expressed as [27]

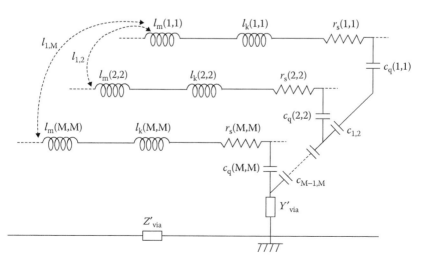

FIGURE 4.10
MTL model of an MWCNT via.

$$\frac{\partial V}{\partial x} = -\left[r_s + j\omega(l_m + l_k) + Z'_{\text{via}} \right] I \qquad (4.30)$$

$$\frac{\partial I}{\partial x} = -\left[(j\omega C + Y'_{\text{via}})^{-1} + (j\omega C_q)^{-1} \right]^{-1} V \qquad (4.31)$$

The inductance l_m is primarily of two types: (1) the via self-magnetic inductance $l_m(i,i)$ that is due to the induced magnetic field by current flowing through an MWCNT shell and (2) the mutual magnetic inductance $l_{i,i+1}$ that originates due to the magnetic field coupling between the adjacent shells. The coupling capacitance $c_{i,i+1}$ arises due to the potential difference between the ith and $(i+1)$th shells in MWCNT.

The MTL model of Figure 4.10 is further simplified to an ESC model of MWCNT bundled via as shown in Figure 4.11. In order to model the parasitics of an MWCNT bundled via, it is necessary to obtain the total number of conducting channels that primarily depends on the number of MWCNTs in a bundle. The number of conducting channels of each MWCNT is obtained using the number of conducting channels of each shell.

The equivalent scattering resistance ($r_{s,\text{MWCNT}}^{\text{Bundle}}$), kinetic inductance ($l_{k,\text{MWCNT}}^{\text{Bundle}}$), and quantum capacitance ($c_{q,\text{MWCNT}}^{\text{Bundle}}$) of the ESC model are primarily dependent on the $N_{\text{Total,MWCNT}}$ of the bundle. The $R_{C,\text{MWCNT}}^{\text{Bundle}}$ consists of a quantum resistance (R_q) and an imperfect metal–nanotube contact resistance (R_{mc}) and can be expressed as

$$R_{C,\text{MWCNT}}^{\text{Bundle}} = \frac{R_q + R_{mc}}{2N_{\text{Total,MWCNT}}} \qquad (4.32)$$

where:

$$R_q = \frac{h}{2e^2} \approx 12.9\,\text{k}\Omega$$

This chapter considers the transmission line model of a pair of CNT bundled TSVs to analyze the crosstalk, power, and delay performance.

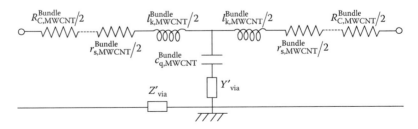

FIGURE 4.11
ESC model of a MWCNT bundled TSV.

4.5 Performance Analysis

This section demonstrates the propagation delay, power dissipation, and crosstalk performance for Cu- and CNT bundled TSVs. The delay and power are analyzed by using a DTL setup, whereas capacitively coupled TSV lines are used to analyze the crosstalk.

4.5.1 Propagation Delay and Power Dissipation Analysis

Propagation delay and power dissipation of SWCNT and MWCNT bundled TSVs are analyzed using a DTL system (Figure 4.12). A resistive driver with a supply voltage $V_{in} = 1$ V is used to drive the via line. The TSV in the DTL system is represented using the equivalent electrical models of bundled SWCNT and bundled MWCNT. The p.u.h. equivalent resistance, inductance, and capacitance are represented as $R'_{ESC} = r_{TSV}$, $L'_{ESC} = l_{TSV}$, and $C'_{ESC} = (1/c_q + 1/c_{TSV})^{-1}$, respectively. The lumped resistance (R_1) in Figure 4.12 is the series combination of imperfect metal–nanotube contact resistance (R_{mc}) and quantum resistance (R_q). The driver resistance and driver capacitance are represented as R_{dr} and C_{dr}, respectively. The via line is terminated by a load capacitance C_L of 10 fF. The quantitative values of driver and via parasitics of SWCNT and MWCNT bundled TSVs are summarized in Table 4.1.

The propagation delay and power dissipation are analyzed for Cu and different number of shells in MWCNTs and SWCNTs in the bundle. Each SWCNT has a diameter of 1 nm, whereas the outermost shell diameters of MWCNTs are in the range of 3.04–7.12 nm. Figure 4.13a and b presents the propagation delay and power dissipation of different TSV filler materials at the TSV height of 30 and 60 µm, respectively. Irrespective of the via height, the overall delay and power dissipation substantially reduce for 10-shell MWCNT bundled TSVs compared with Cu, SWCNT, 4-shell and 8-shell MWCNT bundled TSVs.

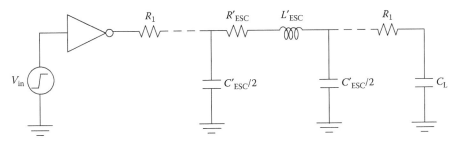

FIGURE 4.12
A DTL system.

TABLE 4.1

Equivalent Parasitics of CNT Bundled TSVs

Via Parasitics	SWB	MWB (4)	MWB (8)	MWB (10)
R'_{ESC} (mΩ/μm)	4.59	0.91	0.42	0.32
L'_{ESC} (pH/μm)	0.73	1.05	1.54	1.72
C'_{ESC} (fF/μm)	0.27	0.17	0.11	0.09
R_{dr} (Ω)	292.36			
C_{dr} (aF)	45.51			

Note: SWB, MWB (4), MWB (8), and MWB (10) represent the bundled SWCNT, 4-shell MWCNT, 8-shell MWCNT, and 10-shell MWCNT bundled TSVs, respectively.

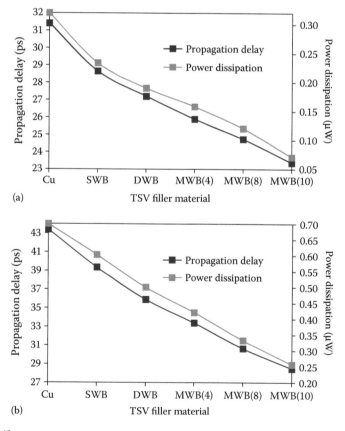

(a)

(b)

FIGURE 4.13
Propagation delay and power dissipation of different TSV filler materials at (a) h_{TSV} = 30 μm and (b) h_{TSV} = 60 μm.

Propagation delay and power dissipation are proportional to the via self-resistance and self-capacitance, which in turn depend on the number of conducting channels. For a fixed via radius, the 10-shell MWCNT bundle has lesser number of CNTs compared to SWCNT bundle, resulting in reduced number of conducting channels. This substantially reduces the c_q while increasing the R'_{ESC}. However, in the 10-shell MWCNT bundled TSV that has larger diameter CNTs, the longer mfp (λ_{mfp}) reduces the overall R'_{ESC} (Table 4.1), which in turn significantly increases the overall conductivity of the bundle. Thus, the cumulative effect of the number of conducting channels and mfp in the 10-shell MWCNT bundled TSV is the overall reduction of R'_{ESC} (increase in conductivity) and c_q. Thus, the overall propagation delay and power dissipation of the 10-shell MWCNT bundled TSV are substantially reduced compared to the Cu-, SWCNT, four-shell MWCNT, and eight-shell MWCNT bundled TSVs.

4.5.2 Crosstalk Analysis

Crosstalk in coupled lines is broadly classified in two categories: (1) functional and (2) dynamic. Under the functional crosstalk category, victim line experiences a voltage spike when the aggressor line switches. However, dynamic crosstalk is observed when aggressor and victim lines switch simultaneously. A change in propagation delay is experienced under dynamic crosstalk when adjacent line (aggressor and victim) switches either in the same direction (in-phase) or in the opposite direction (out-phase) [28]. This section analyzes the in-phase and out-phase crosstalk delays using capacitively coupled TSV lines as shown in Figure 4.14. The TSVs in Figure 4.14 are modeled by the equivalent *RLC* line of different SWCNT, DWCNT, and MWCNT bundles. A CMOS driver with a supply voltage $V_{dd} = 1$ V is used for accurate estimation of crosstalk delay. The TSV lines are terminated by a load capacitance $C_L = 10$ aF.

Using the afore-mentioned setup and equivalent *RLC* model, Figure 4.15 presents the in-phase and out-phase delays for Cu, bundled SWCNT,

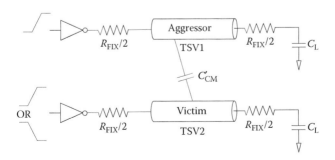

FIGURE 4.14
Capacitively coupled via lines.

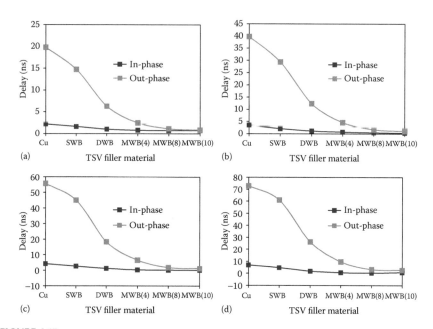

FIGURE 4.15

Crosstalk-induced delays for different TSV filler materials at TSV heights of (a) 50, (b) 100, (c) 150, and (d) 200 µm. SWB, DWB, MWB(4), MWB(8), and MWB(10) represent the bundled SWCNT, DWCNT, and 4-, 8-, and 10-shell MWCNTs, respectively.

DWCNT, and 4-shell, 8-shell, and 10-shell MWCNTs at different TSV heights ranging from 50 to 200 µm with a step size of 50 µm. Irrespective of via heights, it is observed that both the in-phase and out-phase delays are significantly reduced for MWCNT bundle (number of shells = 10) compared to the Cu, SWCNT, DWCNT, and four-shell and eight-shell MWCNT bundled TSVs. The primary reason behind this reduction is the lower C_{CM} value that primarily depends on the number of CNTs in peripheral layers. For a fixed via radius, the number of 10-shell MWCNTs in the periphery is lesser compared to the SWCNT, DWCNT, and four-shell and eight-shell MWCNT bundled TSVs. Therefore, C_{CM} is substantially reduced, which results in minimum crosstalk delay for the bundled TSV having MWCNTs of 10 shells.

Using different TSV filler materials, the percentage reduction of in-phase and out-phase crosstalk delays for the 10-shell MWCNT bundled TSV is presented in Tables 4.2 and 4.3, respectively. Compared to SWCNT bundle, the overall reduction of in-phase and out-phase crosstalk delays of MWCNT bundle (shell = 10) is 96.85% and 85.59%, respectively.

4.5.3 Frequency Response and Bandwidth Analysis

This section analyzes the frequency response and bandwidth of CNT bundled TSVs using the transfer function (TF) of the DTL system (Figure 4.12). The TF accurately takes into account the driver resistance, the driver capacitance,

TABLE 4.2

Percentage Reduction of In-Phase Delay for 10-Shell MWCNT Bundled TSV with Respect to Cu-, SWCNT, DWCNT, and 4- and 8-Shell MWCNT Bundled TSVs

TSV Height (μm)	Reduction in Delay for MWCNT Bundle (Shell = 10) Compared To (%)				
	Cu	SWB	DWB	MWB (Shell = 4)	MWB (Shell = 8)
50	93.6	96.62	91.82	77.08	30.26
100	93.5	96.71	92.02	77.53	31.02
150	94.8	96.79	92.20	77.85	31.52
200	95.9	97.32	93.27	80.92	38.07

Source: Majumder, M. K. et al. *Active and Passive Electronic Components*, vol. 2014, Article ID 524107, pp. 1–7, 2014.

TABLE 4.3

Percentage Reduction of Out-Phase Delay for 10-Shell MWCNT Bundled TSV with Respect to Cu-, SWCNT, DWCNT, and 4- and 8-Shell MWCNT Bundled TSVs

TSV Height (μm)	Reduction in Delay for MWCNT Bundle (Shell = 10) Compared To (%)				
	Cu	SWB	DWB	MWB (Shell = 4)	MWB (Shell = 8)
50	91.4	78.14	60.86	35.98	10.19
100	93.5	85.27	70.12	44.12	10.55
150	95.2	88.61	75.69	49.89	13.49
200	97.9	90.37	78.84	53.67	15.99

Source: Majumder, M. K. et al. *Active and Passive Electronic Components*, vol. 2014, Article ID 524107, pp. 1–7, 2014.

and via parasitics (Table 4.1). Using the transmission matrix parameter for a uniform *RLC* transmission line, the distributed TF of the DTL (Figure 4.12) can be expressed as

$$\text{TF} = \frac{1}{A + sC_L B} \tag{4.33}$$

where $C_L = 10$ fF is the load capacitance and the coefficients A and B can be expressed as

$$A = 1 + s\left[\frac{R'_{\text{ESC}}C'_{\text{ESC}}(nx)^2}{2} + R_{\text{dr}}C_{\text{dr}} + C'_{\text{ESC}}(nx)(R_1 + R_{\text{dr}})\right]$$
$$+ s^2\left[\frac{L'_{\text{ESC}}C'_{\text{ESC}}(nx)^2}{2} + \frac{R'^2_{\text{ESC}}C'^2_{\text{ESC}}(nx)^4}{4!} + \frac{R'_{\text{ESC}}R_{\text{dr}}C'_{\text{ESC}}C_{\text{dr}}(nx)^2}{2}\right. \tag{4.34}$$
$$\left. + \frac{R'_{\text{ESC}}C'^2_{\text{ESC}}(nx)^3(R_1 + R_{\text{dr}})}{3!} + R_1 R_{\text{dr}}C_{\text{dr}}C'_{\text{ESC}}(nx)\right]$$

$$B = \left(2R_1 + R_{dr} + R'_{ESC}(nx)\right) + s\left[\frac{R'_{ESC}C'_{ESC}(nx)^2}{2}(2R_1 + R_{dr})2R_1R_{dr}C_{dr}\right.$$

$$+\frac{R'^2_{ESC}C'_{ESC}(nx)^3}{3!} + L'_{ESC}(nx) + R_1^2C'_{ESC}(nx) + R'_{ESC}R_{dr}C_{dr}(nx)$$

$$\left.+R_1R_{dr}C'_{ESC}(nx)\right] + s^2\left[(2R_1 + R_{dr})\left(\frac{L'_{ESC}C'_{ESC}(nx)^2}{2} + \frac{R'^2_{ESC}C''^2_{ESC}(nx)^4}{4!}\right)\right. \qquad (4.35)$$

$$+R_1R_{dr}R'_{ESC}C'_{ESC}C_{dr}(nx)^2 + \frac{2R'_{ESC}L'_{ESC}C'_{ESC}(nx)^3}{3!} + \frac{R'^3_{ESC}C'^2_{ESC}(nx)^5}{5!}$$

$$\left.+\frac{R_1^2R'_{ESC}C'^2_{ESC}(nx)^3}{3!} + \frac{R'^2_{ESC}R_{dr}C_{dr}C'_{ESC}(nx)^3}{3!} + R_{dr}C_{dr}\left(L'_{ESC} + R_1^2C'_{ESC}\right)(nx)\right]$$

where n and x represent the number of distributed segments and the length of each segment, respectively. The TF of Equation 4.33 is used to obtain the cutoff frequency (f_c) that primarily depends on the via parasitics as presented in Table 4.1. For a fixed via radius of 2.5 μm, Figure 4.16a and b plots the frequency response of DTL at the TSV heights of 30 and 60 μm, respectively. It is observed that the 10-shell MWCNT bundled TSV demonstrates higher f_c compared to the Cu- and SWCNT bundled TSVs. This fact can be explained in terms of dominating via parasitics R'_{ESC} and C'_{ESC}. It leads to form an RC low-pass filter that has a bandwidth close to the cutoff frequency, $f_c = 1/2\pi R'_{ESC}C'_{ESC}$ [29]. For a fixed via height and radius, the time constant $(R'_{ESC}C'_{ESC})$ of the 10-shell MWCNT bundle is smaller compared to the other TSV bundles, which results in higher value of f_c. Therefore, the bandwidth of 10-shell MWCNT bundled TSV is significantly higher compared to the Cu- and SWCNT bundled TSVs.

4.6 Conclusion

This chapter presented accurate analytical models of signal–ground CNT bundled TSVs. The equivalent RLC model of SWCNT and MWCNT bundled TSVs is used to represent the TSV of the DTL system. The TF of the DTL system is obtained by representing the via line with the ESC model of Cu- and CNT bundled TSVs. A second-order TF is derived to analyze the frequency response of CNT bundled TSVs. For a fixed via dimension, the bundled TSV having MWCNTs with more number of shells demonstrates

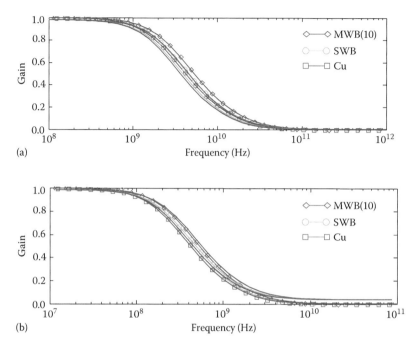

FIGURE 4.16
Frequency response of different TSV filler materials at (a) $h_{TSV} = 30$ μm and (b) $h_{TSV} = 60$ μm.

larger bandwidth compared to the bundled SWCNT and Cu. Moreover, the bundle containing MWCNTs with more number of shells exhibits improved delay and power performance compared to the Cu- and SWCNT bundle-based TSVs.

Using capacitively coupled TSV lines, signal integrity is analyzed for different bundle configurations. It is observed that the overall in-phase and out-phase crosstalk delays are substantially reduced for the bundled TSV having MWCNTs with higher diameters compared to the Cu-based TSV and bundled TSVs having SWCNTs, DWCNTs, and MWCNTs of smaller diameters. It is observed that using an MWCNT bundle (with the number of shells = 10), the overall in-phase delays are reduced by 94.2%, 96.86%, 92.33%, 78.35%, and 32.72% compared to the Cu, bundled SWCNT, DWCNT, eight-shell MWCNT, and eight-shell MWCNT, respectively; similarly, the overall reduction in out-phase delay is 96.5%, 85.89%, 73.38%, 45.92%, and 12.56%, respectively.

Using absolute frequency response, it is observed that the bandwidth of the 10-shell MWCNT bundled TSV is larger compared to the SWCNT bundle- and Cu-based TSVs. Moreover, a 10-shell MWCNT bundle demonstrates a substantial reduction in propagation delay and power dissipation compared to the SWCNT bundle- and Cu-based TSVs.

Multiple Choice Questions

1. Which of the following parameters plays an important role in improving device performance?
   a. Dielectric constant
   b. Threshold voltage
   c. Power supply voltage
   d. Gate-to-drain voltage

2. What is used to increase the speed of operation in device fabrication?
   a. Ceramic gate
   b. Silicon dioxide
   c. Silicon nitride
   d. Polysilicon gate

3. For high-frequency applications, SiO_2 cannot be used due to its
   a. large fringing capacitance
   b. complex fabrication process
   c. Both a and b
   d. None of the above

4. In comparison to the SWCNT bundle- and Cu-based TSVs, the bandwidth of the 10-shell MWCNT bundled TSV is
   a. Equal
   b. Smaller
   c. Larger
   d. None of the above

5. Which of the following crosstalks is observed when aggressor and victim lines switch simultaneously?
   a. Functional crosstalk
   b. Dynamic crosstalk
   c. Both a and b
   d. None of the above

6. Compared to the Cu-, bundled SWCNT, and 4-shell MWCNT and 8-shell MWCNT bundled TSVs, the overall propagation delay and power dissipation of the 10-shell MWCNT bundled TSV are
   a. Equal
   b. Increased
   c. Reduced
   d. None of the above

7. Intrinsic resistance (R_q), in SWCNT, arises due to
   a. The higher nanotube length
   b. Confinement of electrons in a nanowire
   c. Both a and b
   d. None of the above

8. To provide a better transmission and throughput, the number of conducting channels in SWCNTs
   a. Reduces the resistance
   b. Reduces the inductance
   c. Reduces the capacitance
   d. Both a and b

9. The conduction mechanism in CNT is ballistic (lossless) or dissipative due to
   a. Large value of mfp
   b. Large value of tube length
   c. Large value of current density
   d. All of the above

10. The complex effective conductivity of SWCNT bundled TSVs depends on
   a. Intrinsic mutual impedance
   b. Extrinsic self-impedance
   c. Intrinsic self-impedance
   d. Extrinsic mutual impedance

Short Questions

1. What are the advantages of using polymer liners in TSVs over SiO_2 liners?

2. Comment on the metal–electrode contact with CNT.

3. Why SWCNT and MWCNT are preferred over Cu as filler materials?

4. What are the types of resistances in the *RLC* model of SWCNT bundled TSV?

5. What are the types of capacitances in the *RLC* model of SWCNT bundled TSV?

6. What are the types of inductances in the *RLC* model of SWCNT bundled TSV?

7. Give the expressions for effective impedance and admittance of SWCNT bundled TSVs.

8. What are the inductances and capacitances in the MTL model of MWCNT TSVs?

9. How functional and dynamic crosstalks are affected with change in coupling capacitance?

10. List the challenges associated with the implementation of CNT-based TSVs.

Long Questions

1. Explain the fabrication steps to implement CNT-based TSVs.

2. Explain the *RLCG* model of SWCNT bundled TSV.

3. What are the types of resistances, capacitances, and inductances of SWCNT bundled TSV? Provide the closed form of expressions.

4. Describe the MTL model of an MWCNT TSV.

5. Explain the transmission line model of the MWCNT TSV pair.

6. Explain the ESC model of MWCNT bundled TSV.

7. Explain the DTL system for measurement of propagation delay and power dissipation of SWCNT and MWCNT.

8. Compare the propagation delay performance of SWCNT and MWCNT bundled TSVs.

9. Compare the crosstalk performance of SWCNT and MWCNT bundled TSVs.

10. Compare the bandwidths of SWCNT and MWCNT bundled TSVs.

References

1. Noia, B., and Chakrabarty, K. 2011. Pre-bond probing of TSVs in 3D stacked ICs. In *Proceedings of IEEE International Test Conference*, September 20–22, pp. 1–10. Anaheim, CA: IEEE.

2. Pieters, P., and Beyne, E. 2006. 3D wafer level packaging approach towards cost effective low loss high density 3D stacking. In *Proceedings of the IEEE 7th International Conference on Electronic Packaging Technology 2006*, August 26–29, pp. 1–6. Shanghai, People's Republic of China: IEEE.

3. Jones, R. E. 2007. Technology and application of 3D interconnect. In *Proceedings of the IEEE International Conference on Integrated Circuit Design and Technology, 2007*, May 30–June 1, pp. 1–4. Austin, TX.

4. Lo, W. C., Chang, S. M., Chen, Y. H., Ko, J. D., Kuo, T. Y., Chang, H. H., and Shih, Y. C. 2007. 3D chip-to-chip stacking with through silicon interconnect. In *Proceedings of the IEEE International Symposium on VLSI Technology, System and Applications, 2007*, April 23–25, pp. 1–2. Hsinchu, Taiwan.

5. Stan, M. R., Unluer, D., Ghosh, A., and Tseng, F. 2009. Graphene devices, interconnect and circuits–challenges and opportunities. In *Proceedings of the IEEE International Symposium on Circuits and Systems*, May 24–27, pp. 69–72. Taipei, Taiwan.

6. Xu, Y., and Srivastava, A. 2010. A model for carbon nanotube interconnects. *International Journal of Circuit Theory and Applications* 38:559–575.

7. Zhao, W. S., Sun, L., Yin, W. Y., and Guo, Y. X. 2014. Electrothermal modelling and characterisation of submicron through-silicon carbon nanotube bundle vias for three-dimensional ICs. *IET Micro & Nano Letters* 9:123–126.

8. Majumder, M. K., Kaushik, B. K., and Manhas, S. K. 2014. Analysis of delay and dynamic crosstalk in bundled carbon nanotube interconnects. *IEEE Transactions on Electromagnetic Compatibility*. doi:10.1109/TEMC.2014.2318017.

9. Li, H., Xu, C., Srivastava, N., and Banerjee, K. 2009. Carbon nanomaterials for next-generation interconnects and passive, physics, status and prospects. *IEEE Transactions on Electron Devices* 56:1799–1821.

10. Sarto, M. S., and Tamburrano, A. 2010. Single-conductor transmission-line model of multiwall carbon nanotubes. *IEEE Transactions on Nanotechnology* 9:82–92.

11. Gupta, A., Kannan, S., Kim, B. C., Mohammed, F., and Ahn, B. 2010. Development of novel carbon nanotube TSV technology. In *Proceedings of the IEEE 60th Electronic Component and Technology Conference*, pp. 1699–1702. Las Vegas, NV.

12. Xu, C., Li, H., Suaya, R., and Banerjee, K. 2010. Compact AC modeling and performance analysis of through-silicon vias in 3-D ICs. *IEEE Transactions on Electron Devices* 57:3405–3417.

13. Zhao, W. S., Yin, W. Y., and Guo, Y. X. 2012. Electromagnetic compatibility-oriented study on through silicon single-walled carbon nanotube bundle via (TS-SWCNTBV) arrays. *IEEE Transactions on Electromagnetic Compatibility* 54:149–157.

14. Huang, C., Chen, Q., Wu, D., and Wang, Z. 2013. High aspect ratio and low capacitance through-silicon-vias (TSVs) with polymer insulation layers. *Microelectronics Engineering* 104:12–17.

15. Tezcan, S., Duval, F., Philipsen, H., Luhn, O., Soussan, P., and Swinnen, B. 2009. Scalable through silicon via with polymer deep trench isolation for 3D wafer level packaging. In *Proceedings of the IEEE Electronic Components and Technology Conference*, May 26–29, pp. 1159–1164. San Diego, CA.

16. Liang, L., Miao, M., Li, Z., Xu, S., Zhang, Y., and Zhang, X. 2011. 3D modelling and electrical characteristics of through-silicon-via (TSV) in 3D integrated circuits. In *Proceedings of the IEEE 12th International Conference on Electronic Packaging Technology and High Density Packaging*, August 8–11, pp. 1–5. Shanghai, People's Republic of China.

17. Alam, S. M., Jones, R. E., Rauf, S., and Chatterjee, R. 2007. Inter-strata connection characteristics and signal transmission in three-dimensional (3D) integration technology. In *Proceedings of the IEEE 8th International Symposium on Quality Electronic Design*, pp. 580–585. San Jose, CA.

18. Khalil, D. E., Ismail, Y., Khellah, M., Karnik, T., and De, V. 2008. Analytical model for the propagation delay of through silicon vias. In *Proceedings of the IEEE 9th International Symposium on Quality Electronic Design*, pp. 553–556. San Jose, CA.

19. Pak, J. S., Ryu, C., and Kim, J. 2007. Electrical characterization of through silicon via (TSV) depending on structural and material parameters based on 3D full wave simulation. In *Proceedings of the IEEE International Conference on Electronic Materials and Packaging*, pp. 1–6. Daejeon, South Korea.
20. Savidis, I., and Friedman, E. G. 2009. Closed-form expressions of 3-D via resistance, inductance, and capacitance. *IEEE Transactions on Electron Devices* 56(9):1873–1881.
21. Bandyopadhyay, T., Chatterjee, R., Chung, D., Swaminathan, M., and Tummala, R. 2009. Electrical modeling of through silicon and package vias. In *Proceedings IEEE International Conference on 3D System Integration*, pp. 1–8. San Francisco, CA.
22. Naeemi, A., and Meindl, J. D. 2008. Performance modeling for single- and multiwall carbon nanotubes as signal and power interconnects in gigascale systems. *IEEE Transactions on Electron Devices* 55(10):2574–2582.
23. Naeemi, A., and Meindl, J. D. 2006. Compact physical models for multiwall carbon-nanotube interconnects. *IEEE Electron Device Letters* 27(5):338–340.
24. Kannan, S., Gupta, A., Kim, B. C., Mohammed, F., and Ahn, B. 2010. Analysis of carbon nanotube based through silicon vias. In *Proceedings of the 60th IEEE International Conference Electronic Components and Technology*, pp. 51–57. Las Vegas, NV.
25. Kim, B., Gupta, A., Kannan, S., Mohammed, F., and Ahn, B. 2012. Method and model of carbon nanotube based through silicon vias (TSV) for RF applications. US Patent 2012/0306096A1.
26. Gupta, A., Kim, B., Kannan, S., Evana, S. S., and Li, L. 2011. Analysis of CNT based 3D TSV for emerging RF applications. In *Proceedings of the 61st IEEE International Conference on Electronic Components and Technology*, pp. 2056–2059. Lake Buena Vista, FL.
27. Sarto, M. S. and Tamburrano, A. 2009. Multiwall carbon nanotube vias: An effective TL model for EMC oriented analysis. In *Proceedings of the IEEE International Symposium on Electromagnetic Compatibility*, pp. 97–102. Austin, TX.
28. Vittal, A., Chen, L. H., Marek-Sadovvska, M., Wang, K. P., and Yang, S. 1999. Crosstalk in VLSI interconnections. *IEEE Transactions on Computer-Aided Design of Integrated Circuits and Systems* 18(12):1817–1824.
29. Majumder, M. K., Kukkam, N. R., and Kaushik, B. K. 2014. Frequency response and bandwidth analysis of multi-layer graphene nanoribbon and multi-walled carbon nanotube interconnects. *IET Micro & Nano Letters* 9:557–560.
30. Majumder, M. K., Kumari, A., Kaushik, B. K., and Manhas, S. K. 2014. Signal integrity analysis in carbon nanotube based through-silicon via. *Active and Passive Electronic Components*, vol. 2014, Article ID 524107, pp. 1–7.

5

Mixed CNT Bundled Through Silicon Vias

5.1 Introduction

In highly scaled three-dimensional technology, the overall performance and reliability of a chip is strongly dependent on the self and coupling parasitics of through silicon vias (TSVs) and interconnects rather than the transistor logic. Therefore, it is essential to reduce the parasitics, and therefore delay and crosstalk, for efficient performance of future high-speed interconnects and TSVs. The performance of any TSV is primarily dependent on the choice of the filler material used. Recently, mixed carbon nanotubes (CNTs) have emerged as an interesting choice for filler material due to their lower thermal expansion, electromigration, higher mechanical stability, thermal conductivity, and current-carrying capability [1–8].

The performance of Cu- and single-walled CNT (SWCNT) bundled (SWB) TSVs were previously analyzed and compared by Gupta et al. [9] for different via dimensions. The authors proposed an equivalent electrical model by neglecting the imperfect metal–nanotube contact resistance. Xu et al. [10] proposed a compact RLC model of SWB- and multiwalled CNT (MWCNT) bundled TSVs for a fixed number of conducting channels and mean free path. However, Zhao et al. [11] considered the diameter-dependent conducting channels while modeling via parasitics. They analyzed the performance while considering perfect contact resistance ($R_{mc} = 0$) and a fixed value of mean free path ($\lambda_{mfp} = 1$ μm). However, these analyses [10–12] lacked in accuracy and did not analyze the performance thoroughly. Therefore, a more realistic model and analysis is required that can fairly compare the performance of different CNT bundled TSVs.

For a superior performance, researchers have advocated for mixed CNT bundle (MCB) compared to the purely SWB and MWCNT bundles. The crosstalk of a coupled MCB-based TSV can be substantially reduced by maneuvering the placement of MWCNTs and SWCNTs. Several researchers have reported that the outermost shell of a MWCNT carries lesser current compared to an SWCNT with similar diameter [13–15]. Therefore, placing MWCNTs at critical locations in crosstalk-aware designs can significantly improve the signal integrity. Considering these facts, this chapter introduces a novel MCB-based TSV that contains MWCNTs at the periphery and SWCNTs in the core. A comprehensive and accurate analytical model is employed, taking into account the metal–oxide–semiconductor (MOS) effect generated by the presence of TSV in the silicon substrate [10]. The MCB parasitics are accurately modeled by considering the diameter-dependent mean free path and the number of conducting channels [16]. Using capacitively coupled TSV lines, propagation delay, and crosstalk-induced delay are analyzed for SWB- and MCB-based TSVs.

This chapter presents a novel method to reduce crosstalk-induced delay in coupled TSVs containing CNTs as filler materials. A unique structure of MCB-based TSV is proposed in which MWCNTs are placed peripherally to the centrally located SWCNTs. An electrical equivalent model is employed for analyzing these MCB configurations.

This chapter is arranged in six sections including this section. Thereafter, Section 5.2 presents the physical configuration of different CNT bundled TSVs, whereas Section 5.3 describes the equivalent electrical model. Using a coupled driver–TSV–load setup, Section 5.4 analyzes the improvement in signal integrity for peripherally placed MWCNTs in MCB-based TSVs. Finally, Section 5.5 draws the major conclusions of the proposed work.

5.2 Configurations of Mixed CNT Bundled TSVs

The filler material of embedded TSV on the Si substrate is composed of different types of CNT bundles. This section presents the configuration of CNT bundled TSV with associated geometrical and physical parameters. Additionally, the unique configurations of mixed CNT bundled TSVs are also presented in this section.

5.2.1 Physical Configuration of a TSV Pair

The physical configuration of a pair of CNT bundled TSV is shown in Figure 5.1a, and their associated physical and geometrical parameters are shown in Table 5.1. Figure 5.1b and c demonstrate the top cross-sectional view of SWB- and MWCNT bundled TSVs, respectively. The CNT bundle is surrounded by dielectric (usually SiO_2) for dc isolation. Furthermore, the isolation dielectric is surrounded by a depletion region. The thickness of the depletion region primarily depends on the voltage bias condition, interface charge density, material properties of the surrounding Si substrate, and so on [10].

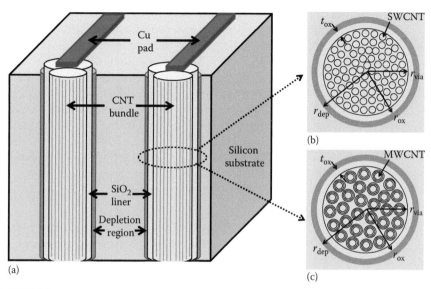

FIGURE 5.1
(a) Physical configuration of a TSV pair, (b) top cross-sectional view of an SWCNT bundled TSV, and (c) top cross-sectional view of an MWCNT bundled TSV.

TABLE 5.1

Geometrical Parameters of TSV

Symbol	Parameters	Value/Range
r_{via}	TSV radius	2.5 μm
r_{dep}	Depletion region radius	3.757 μm
t_{ox}	Oxide thickness	0.5 μm
r_{ox}	Oxide radius	3 μm
h_{tsv}	Via height	30–200 μm
d_{pitch}	TSV pitch	10–30 μm
σ_{Si}	Silicon conductivity	10 S/m
ε_{Si}	Silicon permittivity	$11.68\varepsilon_0$
ε_{ox}	Oxide permittivity	$3.9\varepsilon_0$

5.2.2 Mixed CNT Bundled TSVs

This section presents the cross-sectional spatial arrangement of MCB in the TSV. The strongly conducting SWCNTs are placed at the center of the bundle, where the insignificantly conducting MWCNTs are located at the periphery. Due to interwall interactions within MWCNT, some of the quantum conductance channels are blocked, which in turn redistributes the current nonuniformly over the individual shells within the MWCNT [15].

This phenomenon reduces the overall conductivity of an MWCNT compared to an SWCNT, in which no such interactions occur. Consequently, the outermost shells of MWCNTs substantially reduce the coupling capacitance compared to SWCNTs in coupled TSV lines. Thus, placement of MWCNTs at the peripheral layers is preferred for reduction of the crosstalk in coupled TSVs.

Figure 5.2a–c shows the MCB-I, MCB-II, and MCB-III topologies, in which the four-shell MWCNTs are placed in one, two, and three peripheral layers, respectively. In a similar fashion, 10-shell MWCNTs can be placed at the peripheral layers of the proposed MCB-IV, MCB-V, and MCB-VI topologies. The total numbers of SWCNTs and/or MWCNTs in a bundled TSV can be obtained as [11]

$$N_{\text{CNT}} = \frac{2\pi r_{\text{via}}^2}{\sqrt{3}\left(2r_{\text{CNT}} + \delta\right)^2} \tag{5.1}$$

where $\delta \approx 0.34$ nm is the distance between two neighboring CNTs in the bundle. For a fixed via radius $r_{\text{via}} = 2.5$ μm and SWCNT/MWCNT inner shell radius $r_{\text{CNT}} = 0.5$ nm, Table 5.2 presents the total number of SWCNTs and MWCNTs in different MCB configurations.

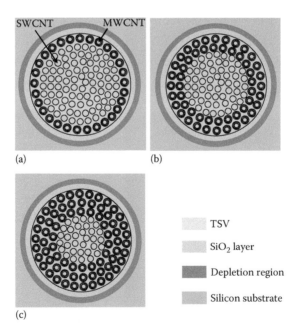

(a)   (b)

(c)

TSV

SiO$_2$ layer

Depletion region

Silicon substrate

FIGURE 5.2
Cross-sectional view of CNT bundled TSVs: (a) MCB-I (one peripheral layer of MWCNTs and the rest of SWCNTs), (b) MCB-II (two peripheral layers of MWCNTs and the rest of SWCNTs), and (c) MCB-III (three peripheral layers of MWCNTs and the rest of SWCNTs).

TABLE 5.2

Number of SWCNTs and MWCNTs in Different MCB-Based TSVs

TSV Bundles	Number of Peripheral Layers of MWCNTs	Number of SWCNTs in the Core	Number of MWCNTs at Peripheral Layers
MCB-I	1 for $N_{\text{shell}} = 4$	12,596,004	4,823
MCB-II	2 for $N_{\text{shell}} = 4$	12,565,352	9,640
MCB-III	3 for $N_{\text{shell}} = 4$	12,534,737	14,451
MCB-IV	1 for $N_{\text{shell}} = 10$	12,554,874	2,317
MCB-V	2 for $N_{\text{shell}} = 10$	12,483,260	4,627
MCB-VI	3 for $N_{\text{shell}} = 10$	12,411,850	6,931

5.3 Modeling of MCB-Based TSVs

This section presents an equivalent electrical model of a pair of TSVs with filler material as different bundled CNTs. Depending on the physical configuration of Figure 5.1a, the equivalent *RLC* model of CNT bundled TSV is shown in Figure 5.3.

The equivalent via resistance (R_{Total}) primarily consists of (1) the quantum or intrinsic resistance (R_q) that is due to quantum confinement of electrons in a nanowire, (2) the imperfect metal–nanotube contact resistance (R_{mc}) that is highly process dependent and exhibits a value of zero to hundreds of kilo-ohms depending on the fabrication process [17], and (3) the scattering resistance (r_{TSV}) that exists for the nanotube lengths exceeding the mean free path

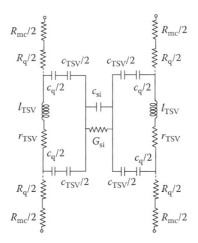

FIGURE 5.3
Equivalent *RLC* model of CNT bundled TSV.

of electrons. Thus, the R_{Total} is the summation of lumped R_q and per-unit height (p.u.h.) r_{TSV}, and can be expressed as

$$R_{Total} = \left(R_{mc} + \frac{R_q + r_{TSV} h_{TSV}}{N_{channel} N_{CNT}} \right)$$

$$= \left[R_{mc} + \frac{h}{2e^2 N_{channel} N_{CNT}} \left(1 + \frac{h_{TSV}}{\lambda_{mfp}} \right) \right] \qquad (5.2)$$

where:

h, e, λ_{mfp}, and h_{TSV} represent Planck's constant, charge of electron, mean free path, and TSV height, respectively

$N_{channel}$ is the number of conducting channels of each shell in MWCNT or each SWCNT in the bundled TSV

The $N_{channel}$ can be modeled using the effect of spin and sublattice degeneracy of carbon atoms and primarily depends on the diameter of each shell in MWCNT and each SWCNT in the bundle [18].

The equivalent inductance (l_{TSV}) appears due to kinetic energy associated with the velocity of electrons in each conducting channel. The l_{TSV} in per unit height can be expressed as

$$l_{TSV} = \frac{h}{4e^2 v_F N_{channel} N_{CNT}} \qquad (5.3)$$

where $v_F \approx 8 \times 10^5$ m/s is the Fermi velocity of CNT and graphene. In the equivalent model, the mutual magnetic inductance is neglected as its magnitude is 3 times smaller than the kinetic inductance [19,20].

The equivalent model of Figure 5.3 comprises (1) the quantum capacitance (c_q) that arises due to the density of electronic states in a quantum wire [19] and (2) the inversion region capacitance (c_{TSV}) that depends on the oxide and depletion region capacitances, c_{ox} and c_{dep}, respectively [11]. The equivalent c_q and c_{TSV} in p.u.h. can be expressed as

$$c_q = \frac{2e^2}{h v_F} N_{channel} N_{CNT} \qquad (5.4)$$

$$c_{TSV} = \left(\frac{1}{c_{ox}} + \frac{1}{c_{dep}} \right)^{-1} = \left[\frac{\ln(r_{ox}/r_{via})}{2\pi\varepsilon_{ox}} + \frac{\ln(r_{dep}/r_{ox})}{2\pi\varepsilon_{Si}} \right]^{-1} \qquad (5.5)$$

where r_{ox} and r_{dep} are the radii of SiO$_2$ and depletion region, respectively. Additionally, the elements c_{Si} and G_{Si} represent the silicon substrate capacitance and conductance, respectively, and can be expressed as

$$c_{Si} = \frac{\pi \varepsilon_{Si}}{\cosh^{-1}(d_{pitch}/2r_{dep})} \tag{5.6}$$

$$G_{Si} = \frac{\sigma_{Si} c_{Si}}{\varepsilon_{Si}} \tag{5.7}$$

where:
   σ_{Si} is the conductivity of the silicon substrate
   d_{pitch} is the distance between two TSVs

5.4 Signal Integrity Analysis of MCB-Based TSVs

This section analyzes the signal integrity of coupled TSV lines with different via heights (h_{TSV}) ranging from 30 to 120 μm. Out of these two lines, one is referred as aggressor and the other one as victim shown in Figure 5.4. The TSV lines are represented by the equivalent single conductor model of either SWB bundle or spatially arranged MCBs.

Hewlett simulation program with integrated circuit emphasis (HSPICE) simulations are performed to demonstrate the crosstalk-induced delay for single and different MCB-based TSVs. A change in propagation delay is experienced under dynamic crosstalk when aggressor and victim switch either in the same direction (in-phase) or in the opposite direction (out-phase). The aggressor and victim lines are terminated by a load capacitance $C_L = 10$ fF. For accurate crosstalk analysis, a complementary MOS driver is used to drive the TSV lines [21–24].

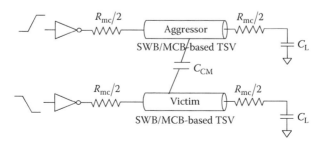

FIGURE 5.4
Coupled via lines switching out-phase.

5.4.1 Worst-Case Crosstalk-Induced Delay for Different TSV Pitches

This section analyzes the worst-case crosstalk-induced delay (out-phase) using coupled TSV lines as shown in Figure 5.4. The out-phase delay is analyzed for SWB and different MCB-based TSVs. The MCBs are considered for two different cases: (1) 4-shell MWCNTs (MCB-I, MCB-II, MCB-III) and (2) 10-shell MWCNTs (MCB-IV, MCB-V, MCB-VI) at the peripheral layers. Using the SWB and MCB-based TSVs, the out-phase delay is analyzed for different pitch distances (d_{pitch}). Figure 5.5a–d demonstrates the out-phase delay for different values of d_{pitch} at TSV heights of 30, 60, 90, and 120 µm, respectively. It is observed that irrespective of via height and pitch distance, the crosstalk-induced delay reduces for an MCB having more number of MWCNTs in peripheral layers. The reduction is substantially higher for peripheral MWCNTs with 10 shells, because it occupies larger cross-sectional area of a bundled TSV. It results in a reduction in quantum capacitance of MCB-VI with MWCNTs

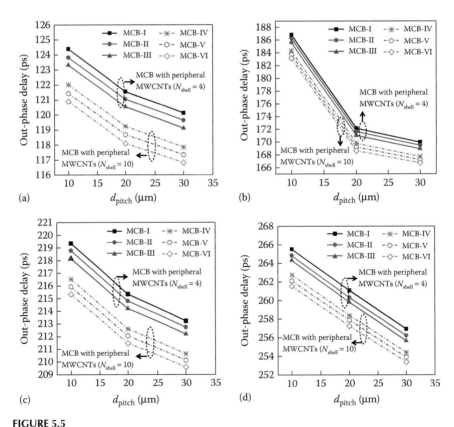

FIGURE 5.5
Out-phase delay of MCB configurations for different pitch distances at TSV heights of (a) 30, (b) 60, (c) 90, and (d) 120 µm [7].

of 10 shells (Table 5.3). Such a configuration reduces the overall crosstalk-induced delay that becomes more prominent for longer TSV heights.

Additionally, from Figure 5.4a–d, it is also observed that irrespective of via height and TSV bundles, the out-phase delay significantly reduces for higher pitch distance due to the dependence of substrate parasitics, c_{Si} and G_{Si}, on the d_{pitch} (Equations 5.6 and 5.7). For higher values of d_{pitch}, the substrate capacitance, c_{Si}, and the substrate conductance, G_{Si}, substantially reduce as presented in Table 5.4. It results in significant reduction of crosstalk-induced delay for more spacing between coupled TSV lines.

In comparison to the conventional SWB, Table 5.5 summarizes the percentage reduction in crosstalk-induced delay for different MCB structures. These

TABLE 5.3

Quantum Capacitance of Different CNT Bundled TSVs

	c_q (nF) for Different h_{TSV} of			
TSV Bundles	**30 μm**	**60 μm**	**90 μm**	**120 μm**
SWB	36.30	72.60	108.90	145.20
MCB-I	24.36	48.73	73.10	97.47
MCB-II	24.34	48.69	73.04	97.38
MCB-III	24.32	48.64	72.97	97.29
MCB-IV	24.30	48.60	72.90	97.20
MCB-V	24.21	48.42	72.63	96.84
MCB-VI	24.12	48.24	72.36	96.48

TABLE 5.4

Substrate Parasitics for Different Pitch Distancees

d_{pitch} (μm)	c_{Si} (fF/μm)	G_{Si} (S/μm)
30	0.16	15.23×10^{-6}
20	0.20	19.21×10^{-6}
10	0.41	39.61×10^{-6}

TABLE 5.5

Percentage Reduction in Out-Phase Delay Using MCB Configurations with Respect to SWCNT Bundled TSVs for $d_{pitch} = 10$ μm

	Reduction in Out-Phase Delay for (%)					
h_{TSV} (μm)	**MCB-I**	**MCB-II**	**MCB-III**	**MCB-IV**	**MCB-V**	**MCB-VI**
30	14.51	14.92	15.26	16.16	16.55	16.93
60	15.36	15.63	15.91	16.51	16.77	17.04
90	16.27	16.49	16.73	17.35	17.57	17.80
120	17.13	17.35	17.51	17.99	18.19	18.38

results are obtained for minimum pitch distance $d_{pitch} = 10$ µm. It is observed that the overall crosstalk-induced delay is reduced by 15.82% and 17.54% for MCB-I and MCB-VI, respectively, compared to the SWB-based TSV.

5.4.2 In-Phase and Propagation Delay for Different TSV Heights

This section presents a comparative analysis of propagation delay and in-phase delay for different CNT bundled TSV configurations. For MCBs with 4-shell and 10-shell peripheral MWCNTs, Figure 5.6a–d presents the propagation delay and crosstalk-induced delay for different TSV heights of 30, 60, 90, and 120 µm, respectively. It is observed that the in-phase delay of MCB-VI is substantially lower compared to other MCB-based TSVs. On an average, the in-phase delay of novel MCB-VI is reduced by 19.73% compared to SWB-based TSV as presented in Table 5.6. In addition to this, the overall in-phase delay is substantially lower compared to the out-phase delay. The reason behind this is the Miller capacitive effect that leads to almost doubling of coupling capacitance. Under out-of-phase transitions, the Miller coupling factor tends to a value of 2 [25].

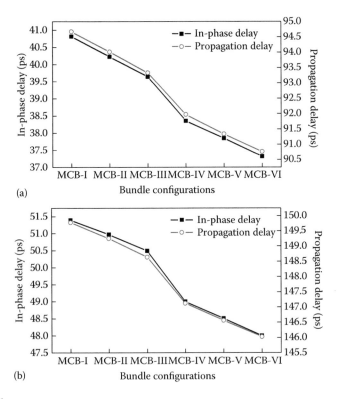

FIGURE 5.6
In-phase and propagation delay of MCB configurations for different TSV heights of (a) 30 µm and (b) 60 µm. *(Continued)*

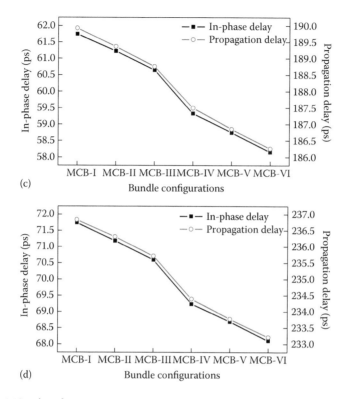

FIGURE 5.6 (Continued)
In-phase and propagation delay of MCB configurations for different TSV heights of (c) 90 μm and (d) 120 μm.

TABLE 5.6

Percentage Reduction in In-Phase Delay Using MCB Configurations with Respect to SWCNT Bundled TSVs for $d_{pitch} = 10$ μm

h_{TSV} (μm)	Reduction in In-Phase Delay for (%)					
	MCB-I	**MCB-II**	**MCB-III**	**MCB-IV**	**MCB-V**	**MCB-VI**
30	12.57	13.83	14.68	16.84	17.94	18.07
60	13.91	14.61	15.42	17.92	18.75	19.61
90	15.43	16.15	16.95	18.75	19.55	20.36
120	16.67	17.31	17.99	19.58	20.21	20.89

A significant reduction in propagation delay is also observed with the novel MCB-VI-based TSV. For different TSV heights, the overall reduction in delay is 14.45% compared to the SWCNT bundled TSV (Table 5.7). The primary reason for this reduction is the higher number of conducting channels of MCB-VI that substantially reduces the via self-resistance and quantum capacitance, thereby reducing the overall propagation delay.

TABLE 5.7

Percentage Reduction in Propagation Delay Using MCB Configurations with Respect to SWCNT Bundled TSVs for $d_{pitch} = 10\ \mu m$

	Reduction in Propagation Delay for (%)					
h_{TSV} (μm)	MCB-I	MCB-II	MCB-III	MCB-IV	MCB-V	MCB-VI
30	10.56	11.09	11.64	13.02	13.54	14.08
60	11.86	12.16	12.52	13.42	13.75	14.27
90	12.82	13.08	13.36	13.94	14.23	14.51
120	13.17	13.37	13.59	14.07	14.49	14.92

5.4.3 Noise Peak Voltage for Different TSV Heights

This section demonstrates the crosstalk-induced positive and negative noise for SWB- and MCB-based TSVs. To obtain the positive peak noise, the victim line remains low and the aggressor line switches from low to high. However, the high–to-low switching on the aggressor line induces a negative peak noise on the victim line. Using SWB-, MCB-III-, and MCB-VI-based TSVs, Figures 5.7 and 5.8 present the crosstalk-induced positive and negative noise at the victim line, respectively. Based on the results, various noise parameters are also calculated to characterize the crosstalk-induced noise. These parameters include the noise peak amplitude (NP), the noise width (NW), and the noise area (NA). The NA is calculated as $NA = 1/2\,NP \times NW$ [26]. At different via heights ranging from 30 to 120 μm, Table 5.8 summarizes the NP, NW, and NA for SWB-, MCB-III-, and MCB-VI-based TSVs.

For MCB-III- and MCB-VI-based TSVs, the NA is substantially improved compared to the SWB-based TSV. This improvement becomes more prominent at longer TSV heights. The NA is primarily dependent on the noise peak voltage and NW. Although the peak voltage of SWB is substantially lower compared to MCB-III and MCB-VI, the MCB-based TSVs demonstrate smaller NW compared to SWB. The NW is primarily dependent on the quantitative value of an equivalent bundle capacitance. The equivalent capacitance of CNT bundled TSVs can be expressed as

$$c_{equ} = \left(\frac{1}{c_q} + \frac{1}{c_{si}} + \frac{1}{c_{TSV}} \right)^{-1} \tag{5.8}$$

The substrate and inversion region capacitances, c_{si} and c_{TSV}, possess similar values for SWB-, MCB-III-, and MCB-VI-based TSVs. For MCB-based TSVs, the quantum capacitance (c_q) is significantly smaller than that of SWB-based TSVs, which exhibit a lower NW. Therefore, the NA of spatially arranged MCB-VI is substantially lower compared to SWB-based TSVs.

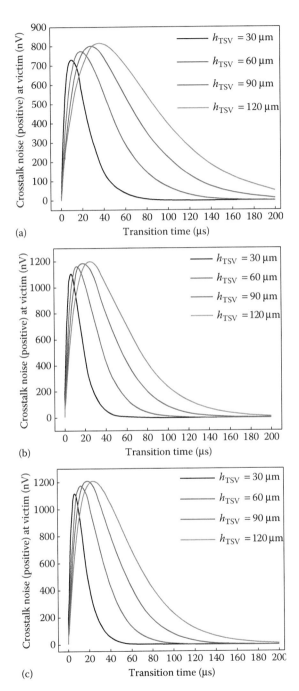

FIGURE 5.7
Crosstalk-induced positive noise at the victim line for (a) SWB-, (b) MCB-III-, and (c) MCB-VI-based TSVs.

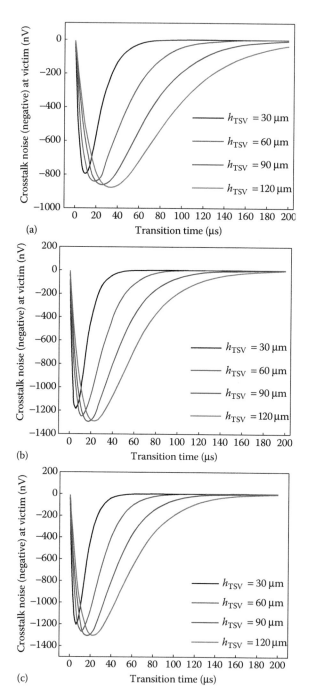

FIGURE 5.8
Crosstalk-induced negative noise at the victim line for (a) SWB-, (b) MCB-III-, and (c) MCB-VI-based TSVs.

TABLE 5.8

Crosstalk-Induced Noise Comparison for SWB- and MCB-Based TSVs

TSV Height (μm)	Noise Peak (nV)			Noise Width (μs)			Noise Area (V ps)		
	SWB	MCB-III	MCB-VI	SWB	MCB-III	MCB-VI	SWB	MCB-III	MCB-VI
30	760.71	1095.26	1117.63	29.53	17.65	16.18	11.23	9.67	9.04
60	778.71	1161.53	1179.52	50.02	32.35	31.79	19.48	18.79	18.75
90	822.16	1175.75	1209.21	78.08	47.05	45.51	32.09	27.65	27.51
120	829.58	1194.46	1211.32	98.57	60.01	58.95	40.89	35.84	35.71

5.5 Summary

This chapter presented unique spatial arrangements of CNTs in a mixed bundled TSV that significantly reduces the capacitive crosstalk in coupled TSV lines. To minimize the effect of capacitive coupling, MWCNTs with different number of shells are placed at the periphery with respect to the centrally located SWCNTs. The analysis demonstrates that a significant reduction in propagation delay and crosstalk-induced delay is obtained for an arrangement in which a larger number of peripheral layers are occupied by MWCNTs. Moreover, the crosstalk reduction is prominently higher for MWCNTs with more number of shells and longer TSVs. Using the proposed MCB arrangement, instead of simple SWBs, with TSV heights ranging from 30 to 120 μm, the average reductions in propagation delay, out-phase delay, and in-phase delay are 14.45%, 17.54%, and 19.73%, respectively.

Multiple Choice Questions

1. For high performance, the placement of MWCNTs in the MCB-based TSV is at
   a. The periphery
   b. The core
   c. There is no use of MWCNTs
   d. None of the above

2. The conductivity of the MWCNT is reduced due to
   a. Increase in conducting channels
   b. Decrease in contact resistance
   c. Both a and b
   d. None of the above

3. When MWCNTs are placed at the periphery, the crosstalk in the device
   a. Reduces
   b. Increases
   c. Not affected
   d. None of the above

4. The advantage of MCB-based TSVs over SWB-based TSVs is
   a. Easy to fabricate
   b. High conductivity
   c. Both a and b
   d. None of the above

5. Crosstalk-induced noise comparison of SWB- and MCB-based TSVs is
   a. SWB=MCB-III=MCB-VI
   b. SWB>MCB-III>MCB-VI
   c. SWB<MCB-III<MCB-VI
   d. None of the above

6. When the radius of the TSV is increased by a factor of 2, the total numbers of SWCNTs and/or MWCNTs in a bundled TSV will
   a. Increase by a factor of 2
   b. Decrease by a factor of 2
   c. Decrease by a factor of 4
   d. None of the above

7. When the radius of the TSV is increased by a factor of 2, the inductance per unit length will
   a. Increase by a factor of 2
   b. Decrease by a factor of 2
   c. Increase by a factor of 4
   d. None of the above

8. As the pitch distance increases, the out-phase delay will
   a. Increase
   b. Decrease
   c. Does not depend on the pitch distance
   d. None of the above

9. As the height of TSV is increased, the in-phase and propagation delay will
   a. Increase
   b. Decrease
   c. Not be affected
   d. None of the above

10. As the NP and NW are increased and decreased by a factor of 2, respectively, the NA would be
    a. Increased by a factor of 2
    b. Decreased by a factor of 2
    c. The same
    d. None of the above

Short Questions

1. Why the overall conductivity of MWCNT is less than the SWCNT?
2. How the crosstalk is reduced with the application of MWCNTs at the periphery?
3. Explain the variation of the NW with an increase in load capacitance.
4. Why the out-phase delay decreases as the pitch distance is increased?
5. Compare the delay performance of MWCNT bundled and MCB-based TSVs.

Long Questions

1. Describe the configuration of MCB-based TSVs.
2. Explain the modeling of the MCB-based TSV with the help of its electrical equivalent diagram and find the expressions for capacitance and resistance?

3. Explain the crosstalk effects in the coupled MCB-based TSVs and find the expression for coupling capacitance assuming the worst case and the best case for crosstalk?

4. Briefly explain the effect of the noise peak in the TSV and the relation of the noise peak with the TSV height.

5. Briefly explain the fabrication process of MWCNT bundled and MCB-based TSVs.

References

1. Majumder, M. K., Kumari, A., Kaushik, B. K., and Manhas, S. K. 2014. Analysis of crosstalk delay using mixed CNT bundle based through silicon vias. In *Proceedings IEEE Radio Frequency Integrated Circuits Symposium*, June 1–3, 441–444. Tampa Bay, FL.

2. Majumder, M. K., Pandya, N. D., Kaushik, B. K., and Manhas, S. K. 2012. Dynamic crosstalk effect in mixed CNT bundle interconnects. *IET Electronics Letters* 48(7):384–385.

3. Majumder, M. K., Kumar, J., Kumar, V. R., and Kaushik, B. K. 2014. Performance analysis for randomly distributed mixed CNT bundle interconnects. *IET Micro & Nano Letters* 9(11):792–796.

4. Majumder, M. K., Das, P. K., and Kaushik, B. K. 2014. Delay and crosstalk reliability issues in mixed MWCNT bundle interconnects. *Microelectronics Reliability, Elsevier* 54(11):2570–2577.

5. Cheung, C. L., Kurtz, A., Park, H., and Lieber, C. M. 2002. Diameter-controlled synthesis of carbon nanotubes. *Journal of Physical Chemistry B* 106(10):2429–2433.

6. Sarto, M. S., and Tamburrano, A. 2006. Electromagnetic analysis of radio-frequency signal propagation along SWCNT bundles. In *Proceedings IEEE International Symposium on Nanotechnology*, pp. 201–204. Cincinnati, OH.

7. Sathyakam, P. U., and Mallick, P. S. 2011. Transient analysis of mixed carbon nanotube bundle interconnects. *Electronics Letters* 47(20):1134–1136.

8. Sathyakam, P. U., Karthikeyan, A., and Mallick, P. S. 2013. Role of semiconducting carbon nanotubes in crosstalk reduction of CNT interconnects. *IEEE Transactions on Nanotechnology* 12(5):662–664.

9. Gupta, A., Kannan, S., Kim, B. C., Mohammed, F., and Ahn, B. 2010. Development of novel carbon nanotube TSV technology. In *Proceedings of the IEEE 60th Electronic Component and Technology Conference*, pp. 1699–1702. Las Vegas, NV.

10. Xu, C., Li, H., Suaya, R., and Banerjee, K. 2010. Compact AC modeling and performance analysis of through-silicon vias in 3-D ICs. *IEEE Transactions on Electron Devices* 57(12):3405–3417.

11. Zhao, W. S., Yin, W. Y., and Guo, Y. X. 2012. Electromagnetic compatibility-oriented study on through silicon single-walled carbon nanotube bundle via (TS-SWCNTBV) arrays. *IEEE Transactions on Electromagnetic Compatibility* 54(1):149–157.

12. Das, P. K., Majumder, M. K., and Kaushik, B. K. 2014. Dynamic crosstalk analysis of mixed multi-walled carbon nanotube bundle interconnects. *IET—The Journal of Engineering.* doi:10.1049/joe.2013.0272.
13. Collins, P. G., Arnold, M. S., and Avouris, Ph. 2001. Engineering carbon nanotubes and nanotube circuits using electrical breakdown. *Science* 292(5517):706–709.
14. Subash, S., Kolar, J., and Chowdhury, M. H. 2013. A new spatially rearranged bundle of mixed carbon nanotubes as VLSI interconnection. *IEEE Transactions on Nanotechnol* 12(1):3–12.
15. Sanvito, S., Kwon, Y. K., Tomanek, D., and Lambert, C. J. 2000. Fractional quantum conductance in carbon nanotubes. *Physical Review Letters* 84(9):1974–1977.
16. Kannan, S., Gupta, A., Kim, B. C., Mohammed, F., and Ahn, B. 2010. Analysis of carbon nanotube based through silicon vias. In *Proceedings of the 60th IEEE International Conference Electronic Components and Technology*, pp. 51–57. Las Vegas, NV.
17. Kumar, V. R., Kaushik, B. K., and Patnaik, A. 2015. Crosstalk noise modeling of multiwall carbon nanotube (MWCNT) interconnects using finite-difference time-domain (FDTD) technique. *Microelectronics Reliability* (Elsevier) 55(1):155–163.
18. Naeemi, A. and Meindl, J. D. 2008. Performance modeling for single- and multi-wall carbon nanotubes as signal and power interconnects in gigascale systems. *IEEE Transactions on Electron Devices* 55(10):2574–2582.
19. Li, H., Yin, W. Y., Banerjee, K., and Mao, J. F. 2008. Circuit modeling and performance analysis of multi-walled carbon nanotube interconnects. *IEEE Transactions on Electron Devices* 55(6):1328–1337.
20. Sarto, M. S., and Tamburrano, A. 2010. Single conductor transmission-line model of multiwall carbon nanotubes. *IEEE Transactions on Nanotechnology* 9(1):82–92.
21. Kaushik, B. K., and Sarkar, S. 2008. Crosstalk analysis for a CMOS-gate-driven coupled interconnects. *IEEE Transactions on Computer-Aided Design of Integrated Circuits and Systems* 27(6):1150–1154.
22. Kaushik, B. K., Sarkar, S., Agarwal, R. P., and Joshi, R. C. 2010. An analytical approach to dynamic crosstalk in coupled interconnects. *Microelectronics Journal* 41(2–3):85–92.
23. Kumar, V. R., Kaushik, B. K., and Patnaik, A. 2014. An accurate model for dynamic crosstalk analysis of CMOS gate driven on-chip interconnects using FDTD method. *Microelectronics Journal* (Elsevier) 45(4):441–448.
24. Kumar, V. R., Kaushik, B. K., and Patnaik, A. 2014. An accurate FDTD model for crosstalk analysis of CMOS-gate-driven coupled RLC interconnects. *IEEE Transactions on Electromagnetic Compatibility* 56(5):1185–1193.
25. Rabaey, J. M. 2002. *Digital Integrated Circuits, A Design Perspective*, 2nd edition. Delhi, India: Prentice Hall.
26. Sahoo, M., and Rahaman, H. 2013. Modeling of crosstalk delay and noise in single-walled carbon nanotube bundle interconnects. In *Proceedings 2013 Annual IEEE India Conference*, December 13–15, pp. 1–6. Mumbai, India.

6

Graphene Nanoribbon-Based
Through Silicon Vias

6.1 Introduction

The use of through silicon via (TSV)-based three-dimensional integrated circuits (3D ICs) has become inevitable in the chip and packaging industry in order to continue meeting the demands of enhanced state-of-the-art electronic products. The TSVs play a major role in bringing out the benefits of 3D integration. Continuous improvements should be carried out in order to maintain TSV-based 3D ICs as a packaging technique that can cater the needs of the current- and the next-generation electronic devices. The properties of a TSV are mainly dependent on the type of filler material used. Currently, Cu is the most commonly used filler material in TSVs [1]. However, graphene-based carbon nanotubes (CNTs) have demonstrated better electrical, thermal, and physical properties compared to Cu [2]. Encouragingly, CNT-based TSVs have shown improved performance in terms of delay, power dissipation, and bandwidth compared to Cu-based TSVs. These improved results suggest that the CNTs have potential to replace Cu as a filler material in future 3D TSVs.

Recently, another graphene-based material, graphene nanoribbon (GNR), has shown improved electrical, thermal, and mechanical properties than Cu. As interconnects, GNRs have provided improved results in terms of delay, power dissipation, and bandwidth compared to Cu [3–7]. It would be a novel concept to observe the results for GNRs as filler materials in TSVs. GNRs exhibit improved performance than Cu as well as CNTs, and thus, it would open a new domain of research to implement GNRs as filler materials in the near future.

Multilayer GNR (MLGNR)-based TSVs significantly improve the conductivity and reduce the propagation delay compared to the Cu-based TSVs at high operating frequencies. This is due to the fact that at higher operating frequencies, contrary to MLGNR, the Cu-based TSVs suffer from the skin effect. Using the COMSOL Multiphysics solver, Hossain et al. [8] analyzed and compared the performance of Cu-, multiwalled CNT (MWCNT)-, and MLGNR-based TSVs. They observed that for a higher current density of 70 A/m², the copper- and CNT-based TSVs exhibit inconsistent and poor performance due to electromigration and coupling incompatibility, respectively, whereas the total energy flux of MLGNR-based TSVs remains consistent and uniform that makes them a better choice. However, the research community still looks forward for an experimental evidence of usage of MLGNRs in TSVs. With the advancement in fabrication technologies and continuous improvements through research and development, it is expected that the issues related to fabrication of MLGNR-based TSVs may be resolved in the near future.

This chapter is arranged in seven sections including this section. To analyze the performance, Section 6.2 provides the novel configuration of GNR-based TSVs. Section 6.3 highlights the fabrication challenges and other limitations that are faced when implementing the GNR-based TSVs. Sections 6.4 and 6.5 present the modeling of GNR-based TSVs with smooth edges and with rough edges, respectively. Section 6.6 deals with the signal integrity analysis of Cu-, SWCNT-, MWCNT-, mixed CNT bundle (MCB)-, and GNR-based TSVs. This analysis demonstrates the performance comparison of various TSVs in terms of propagation delay and crosstalk-induced delay. Finally, Section 6.7 provides a brief summary of this chapter.

6.2 Configurations of GNR-Based TSVs

The physical configuration of an MLGNR-based TSV is shown in Figure 6.1, and its associated physical and geometrical parameters are shown in Table 6.1. The MLGNR in the TSVs are surrounded by dielectric (usually SiO_2) for dc isolation. Furthermore, the isolation dielectric is surrounded by a depletion region. The thickness of the depletion region primarily depends on the voltage bias condition, interface charge density, material properties of the surrounding Si substrate, and so on [9].

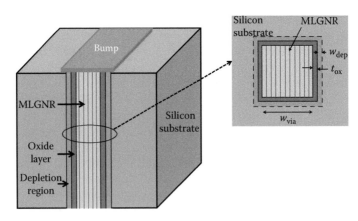

FIGURE 6.1
Physical configuration of an MLGNR-based TSV.

TABLE 6.1

Geometrical Parameters of TSV

Symbol	Parameter	Value/Range
w_{via}	TSV width	2.5 μm
w_{dep}	Depletion region width	3.757 μm
t_{ox}	Oxide thickness	0.5 μm
h_{tsv}	Via height	30–200 μm
d_{pitch}	TSV pitch	10–30 μm
σ_{Si}	Silicon conductivity	10 S/m
ε_{Si}	Silicon permittivity	$11.68\varepsilon_0$
ε_{ox}	Oxide permittivity	$3.9\varepsilon_0$

From the fabrication point of view, it is evident that GNR growth can be more easily controlled than CNT growth because of their planar structure [10,11] that makes them compatible with the conventional lithography techniques [12]. Moreover, appropriate combination of metallic and semiconducting regions of a GNR can be used to develop monolithic structures, that is, both device and TSV can be fabricated on the same graphene layer [10].

6.3 Fabrication Challenges and Limitations

The fabrication challenges associated with CNTs motivate researchers to explore MLGNR-based TSV designing. Most of the electrical and material properties of GNRs are similar to those of CNTs. Additionally, MLGNR-based TSVs possess (1) easier fabrication process due to their rectangular cross-sectional area, (2) higher heat conduction, and (3) swift power delivery. The GNRs have a good compatibility with the conventional lithography technique and are therefore more suitable for horizontal structures such as interconnects than CNTs [13]. However, a vertical integration of GNRs is required for TSVs. The vertical integration demands innovative fabrication techniques in order to simplify the placement of GNRs in TSVs.

It is highly desirable to produce GNRs in a controllable and predictable manner. A procedure involving nanowires as an etch mask to obtain the desired GNRs is shown in Figure 6.2. In this procedure, the graphene is first peeled into layers with the help of a mechanical process on a highly doped wafer of Si with a thermal oxide of 300 nm (Figure 6.2a). In order to remove the organic residue from the substrate, the calcination is done at 300°C. Silicon nanowires of different diameters act as a physical etch mask and are aligned onto the graphene (Figure 6.2b). The Au nanocluster-based vapor–liquid–solid growth mechanism is used to grow the silicon nanowires. A force is produced by submerging the substrate into isopropyl alcohol and blowing nitrogen for drying, in order to establish a good contact between the graphene sheet and the nanowire. With the help of an atomic force microscope, the nanowires are located on the graphene, and thereafter, the graphene layer is selectively etched away with the help of oxygen plasma, leaving GNRs below the nanowire (Figure 6.2c). For the power level of 40 W, the etch time ranges from 20 to 60 s. Initially, the etch is in the vertical direction and results in GNRs that have width in the range of the nanowire diameter. After that, with the help of a lateral etch, the GNR width can be further reduced (Figure 6.2e). Later on, the nanowire mask is removed by sonication. This leaves behind the exposed GNRs on the substrate. (Figure 6.2d and f).

The atomic force microscope image of the graphene flake carrying a nanowire on the top is shown in Figure 6.2g. The application of oxygen plasma etching for about 20 s removed the original graphene flake (Figure 6.2h). Figure 6.2i demonstrates the atomic force microscope image

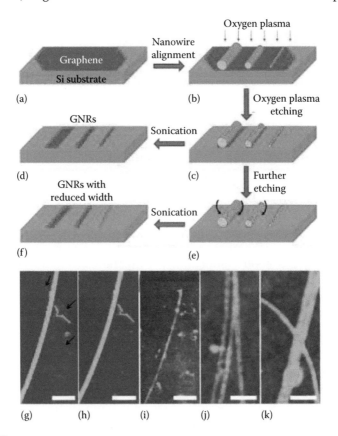

FIGURE 6.2
(a–f) Schematic of fabrication steps to produce GNRs through plasma etching with a nanowire as mask. Atomic force microscopic images of the nanowire on grapahene flake prior to (g) and after (h) plasma etching. (i) Atomic force microscopic image of exposed GNR removing nanowire via sonication, (j) branched GNR from merged nanowire mask, and (k) crossed GNR from crossed nanowire mask. (From Bai, J. et al., *Nano Letters*, 9(5), 2083–2087, 2009. With Permission.)

of the GNR exposed when nanowire is removed by sonication. The pattern of the obtained GNRs is similar to the nanowire. Therefore, different GNR structures such as crossed or bifurcated structures can be obtained from cross nanowires or nanowire bundles (Figure 6.2j and k) [14].

6.4 Modeling of GNR-Based TSVs with Smooth Edges

In general, the performance of MLGNRs is evaluated by means of an electrical equivalent model. This section presents the electrical equivalent models of an MLGNR-based TSV with smooth edges. The multiconductor transmission line (MTL) model of an MLGNR-based TSV is shown in Figure 6.3.

The quantitative values of via parasitics are primarily dependent on the total number of layers (N_{layer}) in MLGNR. The N_{layer} can be expressed as

$$N_{layer} = 1 + \text{Integer}\left(\frac{t}{\delta}\right) \tag{6.1}$$

where t and δ represent the thickness and the interlayer distance, respectively. The value of δ is assumed to be 0.575 and 0.34 nm for doped and neutral MLGNRs, respectively [15]. The number of conducting channels (N_{ch}) of each layer in MLGNR is the sum of total sub-bands that participate in the current conduction. The N_{ch} takes into account the effect of spin and sublattice degeneracy of carbon atoms and primarily depends on the width, Fermi energy (E_f), and temperature (T) [16]:

$$N_{ch} = \sum_{n=0}^{n_C}\left[e^{(E_n - E_f)/kT} + 1\right]^{-1} + \sum_{n=0}^{n_V}\left[e^{(E_n + E_f)/kT} + 1\right]^{-1} \tag{6.2}$$

where:
n_C, n_V, and k represent the number of conduction and valence bands and the Boltzmann constant, respectively
E_n is the lowest/highest energy of nth sub-band in the conduction/valence band

Depending on the current fabrication process, the imperfect metal–MLGNR contact resistance (R_{mc}) has a typical value ranging from 1 to 20 kΩ [17]. Each layer of MLGNR exhibits the lumped quantum resistance (R_q), which is due to the quantum confinement of carriers across the TSV width [18] and can be expressed as

$$R_q^{j,j} = \frac{h}{2e^2 \cdot N_{ch}} \tag{6.3}$$

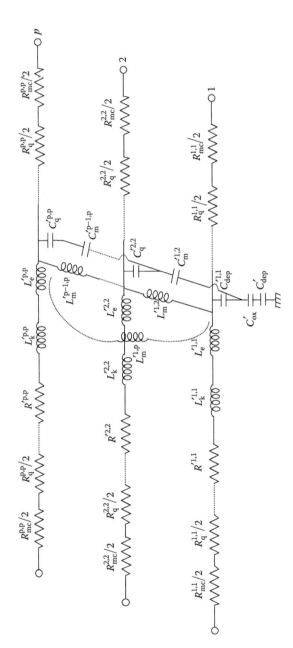

FIGURE 6.3
MTL model of the MLGNR-based TSV.

However, for longer TSVs, scattering resistance $R'^{j,j}$ appears due to the static impurity scattering, defects, line edge roughness (LER) scattering, and so on [19–22]. Thus, $R'^{j,j}$ primarily depends on the effective MFP of electrons (λ_{eff}) and can be expressed as

$$R'^{j,j} = \frac{h}{2e^2 \bullet N_{ch} \bullet \lambda_{eff}} \tag{6.4}$$

Using Matthiessen's rule, the λ_{eff} of each sub-band can be expressed as

$$\frac{1}{\lambda_{eff}} = \frac{1}{\lambda_d} + \frac{1}{\lambda_n} \tag{6.5}$$

where λ_d and λ_n are the mean free path (MFP) due to defects and edge roughness, respectively.

For the MLGNR with smooth edges (no edge roughness), λ_{eff} purely depends on the scattering effects due to defects, λ_d. According to the experimental measurements reported in [12], λ_d is about 1 μm for a single-layer GNR, which is width independent [12,23]. However, in multilayer GNR, the MFP reduces due to the intersheet electron hopping [24]. The λ_d of MLGNR can be extracted by measuring the in-plane conductivity of GNR. Using the in-plane conductivity $G_{sheet} = 0.026$ (μΩ·cm)$^{-1}$ [24], layer spacing of 0.34 nm, and $E_f = 0$ of a neutral MLGNR, the λ_d is extracted as 419 nm by solving the following equation [15]:

$$G_{sheet} = \frac{2q^2}{h} \bullet \frac{\pi \lambda d}{h v_f} \bullet 2k_B T \ln\left[2\cosh\left(\frac{E_f}{2k_B T}\right)\right] \tag{6.6}$$

To increase the conductivity of MLGNR, doping can be used such as AsF$_5$. The in-plane conductivity $G_{sheet} = 0.63$ (μΩ·cm)$^{-1}$ and the carrier concentration $n_p = 4.6 \times 10^{20}$ cm^{-3} are observed for the AsF$_5$-intercalated graphite [25]. Using the simplified tight-binding model, the E_f can be expressed as

$$E_f = h v_f \left(\frac{n_p \cdot \delta}{4\pi}\right)^{1/2} \tag{6.7}$$

The value of δ is considered as 0.575 nm. Using the expressions (6.6) and (6.7), E_f and λ are obtained as 0.6 eV and 1.03 μm, respectively.

Each layer in MLGNR comprises the kinetic inductance ($L_k^{j,j}$) and the quantum capacitance ($C_q^{j,j}$), which represent the mobile charge carrier inertia and the density of electronic states, respectively. The $L_k^{j,j}$ and $C_q^{j,j}$ of any layer j can be expressed as

$$L_k^{j,j} = \frac{L'_{k0}}{2N_{ch}} \tag{6.8}$$

where:

$$L'_{k0} = \frac{h}{2e^2 v_F}$$

$$C'^{j,j}_q = 2C'_{q0} \cdot N_{ch} \qquad (6.9)$$

where:

$$C'_{q0} = \frac{2e^2}{h v_F}$$

where $v_F \approx 8 \times 10^5$ m/s represents the Fermi velocity of carriers in graphene [17,22]. Apart from this, the magnetic inductance (L'_e) of MLGNR is due to the stored energies of carriers in the magnetic field [26]. The $L'^{j,j}_e$ can be expressed as

$$L'^{j,j}_e = \frac{\mu_0 \mu_r d}{w} \qquad (6.10)$$

The interlayer mutual inductance (L'_m) and capacitance (C'_m) are mainly due to the electron tunnel transport phenomenon in adjacent layers [17]. The L'_m and C'_m can be expressed as

$$L'^{(j-1,j)}_m = \frac{\mu_0 \delta}{w} \text{ and } C'^{(j-1,j)}_m = \frac{\varepsilon_0 w}{\delta} \qquad (6.11)$$

The analysis of signal propagation along an MLGNR with N_{layer} leads to the solution of a 2N-dimensional system of differential equations that can be highly time consuming. For this reason, the MTL model of Figure 6.3 is simplified to an equivalent single conductor (ESC) model of Figure 6.4, in which all the layers are assumed to be parallel. The resistance (R_1) is the summation of imperfect metal contact and quantum resistance of MLGNR. The per-unit height (p.u.h.) equivalent resistance (R'_{ESC}), inductance (L'_{ESC}), and capacitance (C'_{ESC}) can be expressed as

$$R'_{ESC} = \frac{R'^{j,j}}{N_{layer}}, L'_{K-ESC} = \frac{L'^{j,j}_k}{N_{layer}} \text{ and } L'_{E-ESC} = L'^{j,j}_e \qquad (6.12)$$

$$C'_{ESC} = \left(\frac{1}{C'_{Q-ESC}} + \frac{1}{C'_{ox} + \frac{1}{C'_{dep}}} \right)^{-1} \qquad (6.13)$$

where:

$$C'_{Q-ESC} = C'^{j,j}_q \bullet N_{layer}$$

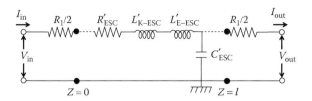

FIGURE 6.4
ESC model of the MLGNR-based TSV.

6.5 Modeling of GNR-Based TSVs with Rough Edges

The fundamental issue in GNR is the presence of edge roughness that substantially reduces the effective MFP. Due to the rough edges, the electrons scatter at the edges, and hence, the effective MFP becomes width dependent. Therefore, the width-dependent MFP has to be incorporated in the model of MLGNR resistance.

The MFP corresponding to the diffusive scattering at the edges is a function of edge backscattering probability (P) and the average distance that traveled by an electron along the length before hitting the edge. The edge scattering MFP for the nth sub-band can be expressed as [27]

$$\lambda_n = \frac{w}{P}\sqrt{\left(\frac{E_F/\Delta E}{n}\right)^2 - 1} \qquad (6.14)$$

where ΔE is the energy gap between the sub-bands. Thus, the scattering resistance can be expressed as

$$R'_{ESC} = \frac{h}{2e^2 N_{layer} N_{ch}} \left[\sum_n \left(\frac{L}{\lambda_{eff}}\right)^{-1}\right]^{-1} \quad \text{for } 0 < P \le 1 \text{ (rough edges)} \quad (6.15)$$

$$R'_{ESC} = \frac{h}{2e^2 N_{layer} N_{ch}} \left(\frac{L}{\lambda_{eff}}\right) \quad \text{for } P = 0 \text{ (smooth edges)} \quad (6.16)$$

Using the above two equations, the scattering resistance of MLGNR for different widths and edge backscattering probabilities is shown in Figure 6.5. The MFP due to defects and Fermi level are assumed to be 419 nm and 0.2 eV, respectively. The thickness and TSV length are considered as 56.9 nm and 100 μm, respectively. From Figure 6.5, it can be observed that the edge roughness increases the resistance by more than 1 order of magnitude, particularly for narrow widths. It is due to the dominating effect of edge scattering at narrow MLGNR widths.

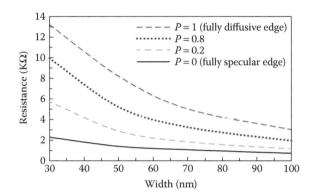

FIGURE 6.5
Resistance of MLGNR for different TSV widths.

6.6 Signal Integrity Analysis of GNR-Based TSVs

The close proximity of TSVs in miniaturized microelectronic devices leads to parasitic capacitive and inductive coupling among them, resulting in crosstalk noise. The crosstalk noise may lead to circuit malfunction, logic failure, change in signal propagation, and unwanted power dissipation. Therefore, accurate modeling of crosstalk noise has emerged as vital design criteria in microelectronics. Propagation delay is also an important parameter in advanced electrical circuits, which determines the speed of the circuits. Improved reliability and speed have become priorities in advanced electronic gadgets. Moreover, it has been previously demonstrated that the filler materials in TSVs play an important role in determining the overall performance of the TSVs. A comparative study of propagation and crosstalk-induced delay for Cu-, SWCNT-, MWCNT-, MCB-, and GNR-based TSVs will help to determine the most suitable material for future TSVs in terms of speed.

An MLGNR-based TSVs, driven by complementary metal–oxide–semiconductor drivers and terminated with capacitive loads, is shown in Figure 6.6. Based on the electrical equivalent model, the performance of MLGNR is compared with that of the other TSV materials such as Cu, SWCNT, MWCNT, and MCB. Moreover, depending on the edge roughness, different values of backscattering probabilities are considered for the comparison purpose. The TSV width, depletion width, and oxide thickness are considered as 2.5, 3.757, and 0.5 μm, respectively. The relative permittivities of silicon and oxide are considered as 11.68 and 3.9, respectively.

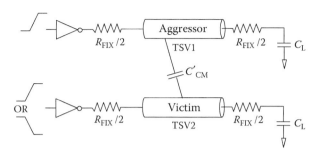

FIGURE 6.6
Structure of coupled TSV lines.

6.6.1 Propagation Delay for Cu-, SWCNT-, MWCNT-, MCB-, and GNR-Based TSVs

The propagation delay for Cu-, SWCNT-, MWCNT-, MCB-, and GNR-based TSVs is obtained for different TSV heights ranging from 30 to 120 μm. The TSV lines switch from the ground to V_{DD}. The delay is analyzed using the industry standard Hewlett simulation program with integrated circuit emphasis (HSPICE) simulator based on the electrical equivalent model. From Figure 6.7, it is observed that the TSVs with MLGNR of $P = 0$ provide the least delay followed by MCB, MLGNR of $P = 0.2$, and MWCNT, SWCNT, Cu, and MLGNR of $P = 1$ TSVs, respectively. Because P signifies the edge roughness in MLGNRs, the higher value of P considerably increases the scattering resistance in MLGNRs, which results in higher delay. The high edge roughness is so detrimental toward delay that the performance of MLGNR of $P = 1$ TSV is outperformed by even the Cu TSV.

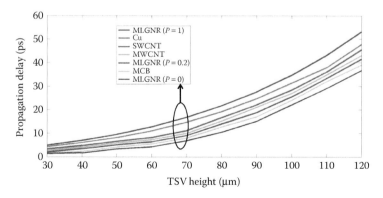

FIGURE 6.7
Propagation delay in different filler material-based TSVs for varying heights.

6.6.2 Crosstalk-Induced Delay for Cu-, SWCNT-, MWCNT-, MCB-, and GNR-Based TSVs

During in-phase crosstalk, both aggressor and victim TSV lines switch from the ground to V_{DD}. For dynamic out-phase crosstalk, the aggressor and victim TSV lines make the transition from the ground to V_{DD} and V_{DD} to the ground, respectively. Comparison of crosstalk-induced in-phase and out-phase propagation delays between coupled TSVs of varying heights from 30 to 120 μm and different filler materials is shown in Figures 6.8 and 6.9, respectively. The Miller coupling capacitance is considered during out-phase switching, whereas it is absent during in-phase switching. Therefore, it can be observed that the out-phase delay is higher than the in-phase delay. Moreover, in case of propagation delay, for both dynamic in-phase and out-phase delays, MLGNR-based TSV (with $P = 0$) provides the least delay followed by MCB, MLGNR with $P = 0.2$, and MWCNT, SWCNT, Cu, and MLGNR with $P = 1$

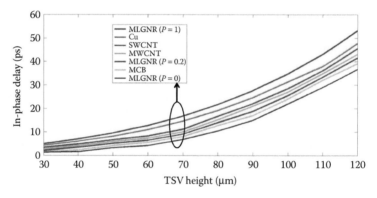

FIGURE 6.8
Comparison of crosstalk-induced in-phase delay between coupled TSVs of different filler materials.

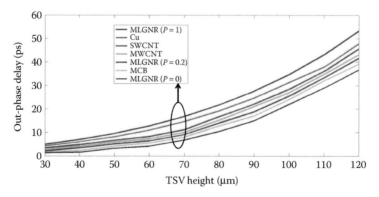

FIGURE 6.9
Comparison of crosstalk-induced out-phase delay between coupled TSVs of different filler materials.

TSVs, respectively. Because these delays depend primarily on line parasitic resistance and capacitance values, it can be understood that the parasitic resistance and capacitance values for MLGNR-based TSV (with $P = 1$) are more due to the dominating effect of LER.

The nonlinear dependence of crosstalk-induced delay on the TSV height can also be observed from Figures 6.8 and 6.9. This is due to the fact that the delay time of the driver–TSV–load system is dependent on the parasitics of the driver, TSV line, and the load. The TSV parasitics are linearly dependent on the TSV height, whereas the driver and load parasitics are not dependent on the TSV height. Also, the driver response to different TSV lines and load parasitics is nonlinear. Therefore, the cumulative effect of the TSV height on the delay time is nonlinear [5].

6.7 Summary

The application of TSVs in 3D ICs is considered as the future of the packaging industry. The filler material in TSVs plays a significant role to determine the overall performance of TSVs. Typically, Cu is used as the filler material; however, recently GNRs have emerged as promising materials that offer better electrical, thermal, and mechanical properties than Cu. This chapter has focused on the novel concept of the application of GNRs as filler materials in TSVs. The physical configuration of GNR-based TSVs is discussed. The important aspect of the fabrication of GNRs along with the challenges and limitations is also highlighted. In order to compare the GNR-based TSV performance with other TSV filler materials such as Cu and CNTs, the electrical equivalent circuit based on the ESC model is provided for both smooth and rough edged GNRs. The signal integrity analyses demonstrate that the purely smooth edge GNR-based TSVs provide improved results in terms of propagation and comprehensive crosstalk-induced delays compared to other filler materials used in TSVs. However, with the increase in GNR edge roughness, the performance of GNR-based TSVs degrades, even falling behind Cu-based TSVs in some cases. Therefore, it is important to focus on the fabrication of smooth edge-based GNRs for their application in future TSVs.

Multiple Choice Questions

1. In TSVs, the overall performance depends on
   a. Pitch distance
   b. Filler material
   c. Crosstalk noise
   d. All of the above

2. If the edge roughness is high, then what happens to the scattering resistance?
   a. Decreases
   b. Increases
   c. Unchanged
   d. None of the above

3. In GNR, the effective MFP reduces due to
   a. Edge roughness
   b. Scattering
   c. Both a and b
   d. None of the above

4. The thickness of the depletion region in MLGNR depends on
   a. Voltage bias condition
   b. Interface charge density
   c. Material properties of the surrounding Si substrate
   d. All of the above

5. The growth of GNR is more controllable than CNT due to
   a. Easier fabrication of GNR
   b. Planner structure of GNR
   c. Material properties of GNR
   d. None of the above

6. In MLGNR, due to edge roughness, there is increase in
   a. Capacitance
   b. Resistance
   c. Inductance
   d. None of the above

7. Which of the following provides the least delay?
   a. MLGNR-based TSV with specular edge
   b. MLGNR-based TSV with diffusive edge
   c. MWCNT
   d. SWCNT

8. The conducting channel in an MLGNR changes with its
   a. Width
   b. Length
   c. Mean free path
   d. Both a and b

9. The energy gap between the sub-bands of an MLGNR primarily depends on the
   a. Width
   b. Thickness
   c. Height
   d. None of the above

10. MLGNR-based TSVs provide better performance than Cu-based TSVs
    a. At lower heights
    b. At higher heights
    c. At all heights
    d. None of the above

Short Questions

1. Explain the crosstalk noise effects in GNR-based TSVs.
2. Write a short note on the propagation delay in GNR-based TSVs.
3. What are the similarities and differences between the electrical and material properties of GNR and CNT?
4. Explain the modeling of GNR-based TSVs with rough edges.
5. Why the parasitic resistance effect is more in fully diffusive MLGNR-based TSVs?
6. Why graphene-based TSVs have demonstrated better performance compared to Cu-based TSVs?
7. What are the basic challenges associated with CNT that can be overcome by MLGNR-based TSVs?
8. Write down the basic steps associated with the fabrication to produce GNRs through plasma etch with a nanowire as mask.

Long Questions

1. Explain in detail the GNR-based TSVs with its fabrication limitations.
2. Explain the fabrication steps used in the production of GNR through plasma etch through nanowire mask.
3. Write a short note on the modeling of GNR-based TSVs with smooth edges and derive the expression for conductivity.

4. Explain the methods to reduce the edge roughness of an MLGNR-based TSV.

5. Compare the different technological-based TSVs based on the crosstalk delay and performance and also explain the variation in propagation delay for different filler materials based on the height variation?

References

1. Kim, J., Pak, J. S., Cho, J. et al. 2011. High frequency scalable electrical model and analysis of a through silicon via (TSV). *IEEE Transactions on Components, Packaging and Manufacturing Technology* 1(2):181–195.

2. Xu, C. et al. 2009. Compact AC modeling and analysis of Cu, W, and CNT based through-silicon vias (TSVs) in 3-D ICs. In *Proceedings of the IEEE International Electron Device Meeting (IEDM 09)*, December, pp. 521–524. Baltimore, MD: IEEE Press.

3. Kumar, V. R., Majumder, M. K., and Kaushik, B. K. 2014. Graphene based on-chip interconnects and TSVs—Prospects and Challenges. *IEEE Nanotechnology Magazine* 8(4):14–20.

4. Kumar, V. R., Majumder, M. K., Kukkam, N. R., and Kaushik, B. K. 2015. Time and frequency domain analysis of MLGNR interconnects. *IEEE Transactions on Nanotechnology* 14(3):484–492.

5. Kumar, V. R., Majumder, M. K., Alam, A., Kukkam, N. R., and Kaushik, B. K. 2015. Stability and delay analysis of multi-layered GNR and multi-walled CNT interconnects. *Journal of Computational Electronics* 14(2):611–618.

6. Sarto, M. S., and Tamburrano, A. 2010. Comparative analysis of TL models for multilayer graphene nanoribbon and multiwall carbon nanotube interconnects. In *Proceedings IEEE International Symposium on Electromagnetic Compatibility*, July, pp. 212–217. Fort Lauderdale, FL.

7. Murali, R., Brenner, K., Yang, Y., Beck, T., and Meindl, J. D. 2009. Resistivity of graphene nanoribbon interconnects. *IEEE Electron Device Letters* 30(6):611–613.

8. Hossain, N. M., Hossain, M., Bin Yousuf, A. H., and Chowdhury, M. H. 2013. Thermal aware graphene based through silicon via design for 3D IC. In *Proceedings IEEE International 3D Systems Integration Conference*, October, pp. 1–4. San Francisco, CA.

9. Kaushik, B. K., Majumder, M. K., and Kumar, V. R. 2014. Carbon nanotube based 3-D interconnects—A reality or a distant dream. *IEEE Circuits and Systems Magazine* 14(4):16–35.

10. Stan, M. R., Unluer, D., Ghosh, A., and Tseng, F. 2009. Graphene devices, interconnect and circuits–Challenges and opportunities. In *Proceedings of the IEEE International Symposium on Circuits and Systems*, May, pp. 69–72. Taipei, Republic of China.

11. Avouris, P. 2010. Graphene: Electronic and photonic properties and devices. *Nano Letters* 10(11):4285–4294.

12. Berger, C., Song, Z., and Lietal, X. 2006. Electronic confinement and coherence in patterned epitaxial graphene. *Science* 312(5777):1191–1196.

13. Kara, M. H., Rahim, N. A. A., Mahmood, M. R., and Awang, Z. 2013. Fabrication and characterization of GNR transmission lines for MMIC applications. In *Proceedings of the 2013 IEEE International RF and Microwave Conference*, December, pp. 42–46. Penang, Malaysia.

14. Bai, J., Duan, X., and Huang, Y. 2009. Rational fabrication of graphene nanoribbons using a nanowire etch mask. *Nano Letters* 9(5):2083–2087.

15. Xu, C., Li, H., and Banerjee, K. 2009. Modeling, analysis, and design of graphene nano-ribbon interconnects. *IEEE Transactions on Electron Devices* 56(8):1567–1578.

16. Nasiri, S. H., Faez, R., and Moravvej-Farshi, M. K. 2012. Compact formulae for number of conduction channels in various types of grapheme nanoribbons at various temperatures. *Modern Physics Letters B* 26(1):1150004-1–115004-5.

17. Cui, J. P., Zhao, W. S., and Yin, W. Y. 2012. Signal transmission analysis of multilayer graphene nano-ribbon (MLGNR) interconnects. *IEEE Transactions on Electromagnetic Compatibility* 54(1):126–132.

18. Chen, Z., Lin, Y., Rooks, M. J., and Avouris, P. 2007. Graphene nano-ribbon electronics. *Physica E: Low-dimensional Systems and Nanostructures* 40(2):228–232.

19. Hwang, E. H., Adam, S., and Sarma, S. D. 2007. Carrier transport in two-dimensional graphene layers. *Physical Review Letters* 98(18):186806-1–186806-4.

20. Yan, J., Zhang, Y., Kim, P., and Pinczuk, A. 2007. Electric field effect tuning of electron-phonon coupling in graphene. *Physical Review Letters* 98(16):166802-1–166802-4.

21. Areshkin, D. A., Gunlycke, D., and White, C. T. 2007. Ballistic transport in graphene nanostrips in the presence of disorder: Importance of edge effects. *Nano Letters* 7(1):204–210.

22. Tan, Y. W., Zhang, Y., Bolotin, K., Zhao, Y., Adam, S., Hwang, E. H., Sarma, S. D., Stormer, H. L., and Kim, P. 2007. Measurement of scattering rate and minimum conductivity in graphene. *Physical Review Letters* 99(24):246803-1–246803-4.

23. Bolotin, K. I., Sikes, K. J., Hone, J., Stormer, H. L. and Kim, P. 2008. Temperature dependent transport in suspended grapheme. *Physical Review Letters* 101(9):096802.

24. Benedict, L. X., Crespi, V. H., Louie, S. G., and Cohen, M. L. 1995. Static conductivity and superconductivity of carbon nanotubes—Relations between tubes and sheets. *Physical Review. B Condensed Matter* 52(20):14935–14940.

25. Hanlon, L. R., Falardeau, E. R., and Fischer, J. E. 1977. Metallic reflectance of AsF5-graphite intercalation compounds. *Solid State Communications* 24(5):377–381.

26. Chen, Z., Lin, Y., Rooks, M. J., and Avouris, P. 2007. Graphene nano-ribbon electronics. *Physica E: Low-dimensional Systems and Nanostructures* 40(2): 228–232.

27. Naeemi, A., and Meindl, J. D. 2007. Conductance modeling for graphene nanoribbon (GNR) interconnects. *IEEE Electron Device Letters* 28(5): 428–431.

7

Liners in Through Silicon Vias

7.1 Introduction

The development of a reliable three-dimensional (3D) integrated system is largely dependent on the choice of liner materials used in through silicon vias (TSVs). Silicon dioxide (SiO_2) is the most commonly used liner material for TSVs. The structure of TSVs is composed of Cu, silicon substrate, and silicon dioxide. Due to the large coefficient of thermal expansion (CTE) mismatch of the silicon substrate and Cu high mechanical stress is induced around Cu. This reveals the reliability issues such as high tensile stress, time-dependent dielectric breakdown failure, and drift of Cu atoms. In addition, due to the high relative permittivity of SiO_2 liner, the liner capacitance becomes high, which degrades the TSV performance. An improved performance can be obtained for a lower value of liner capacitance [1–6]. The liner capacitance is primarily dependent on the dimensions and dielectric constant of the liner material. The silicon dioxide (SiO_2) generates a capacitance of 50 fF-1 pF, with large relative dielectric constant and small thickness (100 nm–1 μm). To reduce the stress effects and liner capacitance offered by SiO_2, researchers proposed the low-k dielectric materials as linear materials. Considering these facts, this chapter analyzes the signal integrity for different liner materials such as SiO_2, polypropylene carbonate (PPC), benzocyclobutene (BCB), and polyimide. The structure of TSVs with polymer liners is shown in Figure 7.1, in which the dies are stacked in a wedding cake style. Using a pair of single-walled carbon nanotube (SWCNT) bundled TSV, the impact of the liner materials on propagation delay and crosstalk-induced delay is analyzed for different via heights and radii.

The chapter is organized as follows: Section 7.1 introduces the recent research scenario and describes briefly the work carried out. Section 7.2 discusses different types of liner materials and their impact on performance, whereas Section 7.3 presents fabrication-related challenges. Section 7.4 demonstrates the modeling of CNT bundled TSVs with liner materials. Using SiO_2 and different polymer as liner materials, Section 7.5 demonstrates the propagation delay and crosstalk-induced delay. Finally, Section 7.6 draws a brief summary.

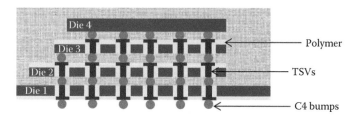

FIGURE 7.1
Structure of TSVs with polymer liners.

7.2 Types of Liners

Several researchers [7–14] reported different electrical and mechanical reliability problems of SiO_2 liners. Recently, Hsin et al. [7] have reported the problem of deposition of uniform thin SiO_2 layer on the sidewalls of deep reactive ion etching (DRIE) vias. Due to the inherent scallops on the sidewalls of TSVs, stress concentration at the scallop ridges results in failure of ultrathin barrier layers and seed layers, current leakage, degradation of electrical performance, and so on [8,9]. Few mechanical reliability problems [10,11], such as die cracks or breakage [12] and interfacial layer contamination [13], come into play because of the large differences in the CTEs of the SiO_2 and the CNT bundles. It largely affects the performance and tolerance of the 3D integrated circuits (ICs) [14]. Therefore, in order to improve the reliability, polymer liner material was introduced to replace SiO_2. The relative permittivity of different TSV liner materials is shown in Table 7.1. Certain features such as large and uniform thickness and low elastic modulus allow the polymers to be suitable materials as TSV liners, which results in improved electrical performance [15]. Current leakage can be avoided by eliminating the discontinuity of barriers using large and uniform thickness of polymer liners. In addition, due to the low elastic modulus, the polymer liner can act as a buffer layer between the silicon substrate and metal plugs to avoid the thermal stresses [16].

TABLE 7.1

Relative Permittivity of Different Liner Materials

Liner Material	Relative Permittivity
Silicon dioxide (SiO_2)	3.9
Polyimide	3.5
PPC	2.9
BCB	2.65

7.3 Fabrication of TSVs with Polymer Liner

The fabrication of TSVs with a polymer liner as an insulating layer is discussed in this section. The schematic of fabrication process flow of TSV is shown in Figure 7.2 [17].

A temporary glue is used to bond the wafer on a glass carrier, and the wafer is then thinned down with the help of rough and fine grinding (Figure 7.2a). Annular patterns are then defined using litho process. Using DRIE, annular trenches are etched for TSVs (Figure 7.2b). Next, the resist is stripped and the dielectric polymer is coated (Figure 7.2c). It is a challenge to completely fill the trenches with polymer. Si surface is exposed by etching the polymer, whereas the trenches are kept filled (Figure 7.2d). A polymer layer is again coated of the wafer surface and is patterned to expose the pillar top of center Si (Figure 7.2e). Later on, the DRIE process is used to etch the centre Si pillar till the premetal dielectric (PMD) layer is reached. Diluted hydrofluoric solution is then used to wet etch the PMD

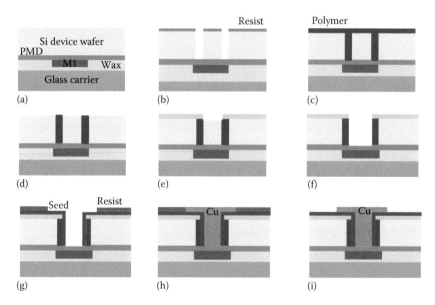

FIGURE 7.2
Schematic process flow of TSV: (a) Bonding wafer on glass carrier using temporary glue. (b) Annular trenches for TSVs using DRIE. (c) Stripping of resist and coating of polymer. (d) Etching the polymer to expose Si surface. (e) A polymer layer coating and patterning to expose the pillar top of center Si. (f) Center Si pillar and PMD etching. (g) Seed layer deposition and patterning. (h) Bottom-up TSV fill and plating of routing lines. (i) Removing a seed layer and plating photoresist. (Reproduced with permission from Tezcan, D. S. et al., *Proceedings of the 59th Electronic Components and Technology Conference (ECTC 2009)*, San Diego, CA, 2009.)

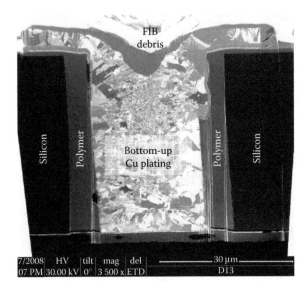

FIGURE 7.3
FIB cross section of the TSV fabricated. (Reproduced with permission from Tezcan, D. S. et al., *Proceedings of the 59th Electronic Components and Technology Conference (ECTC 2009)*, San Diego, CA, 2009.)

layer to expose the landing metal pad (Figure 7.2f). Thereafter, the seed layer of Ti/Cu is deposited on the wafer by sputtering. The last litho step for electroplating Cu layer is performed (Figure 7.2g), and seed repair is optionally done on some wafers. With the help of plating bath, a bottom-up TSV fill along with plating of routing lines on the wafer field is performed (Figure 7.2h). Finally, the seed layer of Ti/Cu and the plating photoresist are removed (Figure 7.2i). A focused ion beam (FIB) cross section of the TSV fabricated is shown in Figure 7.3, in which the bottom-up Cu plating and polymer liner are illustrated clearly.

7.3.1 Polymer Deep Trench Filling

An inner metallization between the electrical feed through and the silicon substrate is needed for dielectric isolation in TSVs. Utilizing thick polymer improves the electrical performance by reducing the capacitive coupling. Moreover, it also improves the reliability by absorbing the stress generated due to CTE mismatch between the Si substrate and the TSV filler material [18]. A simple process to fill deep annular trenches with spin-on polymer through spin coating technique is discussed here. With the help of spray coating and spin coating techniques, the polymer is dispensed on the surface of the wafer to fill the annular trenches and create isolation. A careful spin coating of the polymer is required for successful filling of all trenches. This process is mainly dependent on the following factors [2]:

1. *Surface of wafer on which the deposition of polymer takes place*: It is of ground Si that has a higher roughness than a polished Si wafer. Therefore, wetting issues arise when polymer is dispensed on the wafer. Such issues can be removed by using prewetting agents.

2. *Viscosity of polymer*: A high viscous material does not fill the cavity with the same efficiency as a liquid material. Therefore, it is important to consider the polymer's chemical and physical properties.

3. *TSV density on wafer*: If the number of TSVs on the wafer is higher, it means that more trenches are required to be filled. More numbers of trenches hamper the process of proper cavity filling.

7.4 Modeling of CNT Bundled TSV with SiO$_2$ Liner

The embedded TSVs on the Si substrate primarily have a CNT bundle as a filler material. The physical configuration of a pair of CNT bundled TSVs is shown in Figure 7.4. The TSV is surrounded by a dielectric for dc isolation. Furthermore, the isolation dielectric is surrounded by a depletion region (Figure 7.4). The thickness of this depletion region is primarily dependent on the applied bias voltage, material properties, interface charge density, and so on. Depending on the physical configuration of Figure 7.4, the equivalent electrical model of SWCNT bundled TSV is shown in Figure 7.5.

The overall TSV resistance primarily consists of quantum or intrinsic resistance that is due to the quantum confinement of electrons in a nanowire [19], imperfect metal–nanotube contact resistance (R_{mc}) that primarily depends on the fabrication process [20], and scattering resistance that arises due to the static impurity scattering, defects, line edge roughness scattering, acoustic

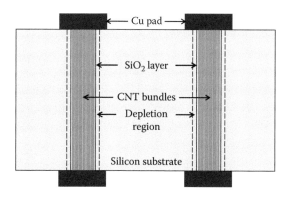

FIGURE 7.4
Physical configuration of a pair of SWCNT bundled TSVs.

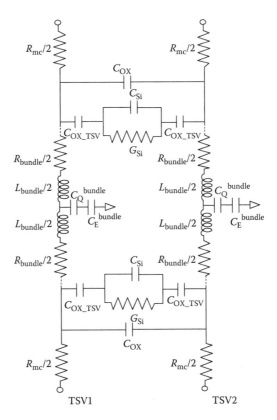

FIGURE 7.5
Equivalent electrical model of a pair of SWCNT bundled TSVs.

phonon scattering, and so on. The equivalent scattering resistance (R_{bundle}) of the bundled TSV can be expressed as [21,22]

$$R_{bundle} = \frac{h}{2e^2\lambda_{mfp}N_{total}} \tag{7.1}$$

where:
  h and e represent Planck's constant and the charge of an electron, respectively
  N_{total} is the total number of conducting channels and can be expressed as

$$N_{total} = N_{channel} \times N_{CNT} \tag{7.2}$$

where:
  $N_{channel}$ is the number of conducting channels of each SWCNT in a bundle [19]
  N_{CNT} represents the total numbers of SWCNTs in a bundled TSV and can be expressed as [22]

$$N_{CNT} = \frac{2\pi r_{via}^2}{\sqrt{3}(D+\delta)} \qquad (7.3)$$

where:

r_{via} and δ (~0.34 nm) are the via radius and the distance between neighboring SWCNTs in the bundle, respectively

The conduction mechanism in CNT is ballistic or dissipative due to the long mean free path (mfp) in the range of micrometers. The diameter-dependent mfp can be expressed as

$$\lambda_{mfp} = \frac{1000D}{(T/T_i)-2} \qquad (7.4)$$

where:

T_i and T represent the temperatures that are equal to 100 and 300 K, respectively [21]

D is the diameter of each SWCNT in the bundled TSV

The equivalent bundle inductance (L_{bundle}) consists of (1) kinetic inductance (L_K) that originates from the kinetic energy of the electrons in each conducting channel and (2) magnetic inductance (L_M) that represents the magnetic field induced by the current flowing through a nanotube [21]. The equivalent L_{bundle} can be expressed as

$$L_{bundle} = \frac{L_K}{2N_{total}} + L_M \qquad (7.5)$$

$$L_K = \frac{h}{2e^2 v_F} \text{ and } L_M = \frac{\mu}{2\pi} \ln\left(\frac{y}{D}\right) \qquad (7.6)$$

where y and v_F represent the distance of CNT bundle from the ground plane and the Fermi velocity of CNT (~8 × 10^5 m/s), respectively.

The equivalent via self-capacitance primarily comprises (1) quantum capacitance (C_Q^{Bundle}) and (2) parallel plate electrostatic capacitance (C_E^{Bundle}). The C_Q^{Bundle} and C_E^{Bundle} can be expressed as [21,23]

$$C_Q^{Bundle} = C_Q \times 2N_{total} \qquad (7.7)$$

where:

$$C_Q = \frac{2e^2}{hv_F}$$

$$C_E^{Bundle} = 2C_{En} + \frac{(n_w-2)}{2}C_{Ef} + \frac{3(n_H-2)}{5}C_{En} \qquad (7.8)$$

where C_{En} and C_{Ef} are the parallel plate capacitances of isolated SWCNT with respect to near and far neighboring TSVs, respectively [23], and can be expressed as

$$C_{En} = \frac{2\pi\varepsilon_0\varepsilon_r}{\ln(2r_{via}/D)} \text{ and } C_{Ef} = \frac{2\pi\varepsilon_0\varepsilon_r}{\ln(4r_{via}/D)} \tag{7.9}$$

For a circular TSV, $n_w = n_H = \sqrt{N_{CNT}}$. The metal–oxide–semiconductor (MOS) capacitance primarily consists of (1) C_{OX} that represents the oxide capacitance between two TSVs and (2) C_{OX_TSV} that appears between the via and the Si substrate [23]. These two capacitances can be expressed as

$$C_{OX\_TSV} = \frac{4\varepsilon_0\varepsilon_r H_{TSV}(r_{via} - t_{ox})}{t_{ox}} \tag{7.10}$$

$$C_{OX} = \left[\frac{2}{C_{OX\_TSV}} + \left(\frac{\varepsilon_0\varepsilon_r A}{d_{pitch}}\right)^{-1}\right]^{-1} \tag{7.11}$$

where:

$$A = \pi r_{via} H_{TSV}$$

where H_{TSV}, d_{pitch}, and t_{ox} are the via height, the center-to-center distance between two TSVs, and the oxide thickness, respectively.

The capacitance and conductance of the lossy Si substrate are represented as C_{Si} and G_{Si}, respectively. These components are primarily dependent on the via radius and d_{pitch}, and can be expressed as [21]

$$C_{Si} = \frac{\varepsilon_0\varepsilon_r A}{d_{pitch}} \tag{7.12}$$

$$G_{Si} = \frac{\pi\sigma}{\ln\left[\frac{d_{pitch}}{2r_{via}} + \sqrt{\left(\frac{d_{pitch}}{2r_{via}}\right)^2 - 1}\right]} \tag{7.13}$$

where $\sigma = 0.1$ (Ω.cm)$^{-1}$ represents the conductivity of the silicon substrate. The quantitative values of the aforementioned via self-parasitics are summarized in Table 7.2. For different bundle aspect ratios (ARs) of 1.2:1 and 5:1, the parasitic values are obtained for 45,000, 37,522, 18,761, and 100 numbers of SWCNTs in a CNT bundled TSV.

TABLE 7.2

Via Parasitics for SWCNT Bundled TSVs

Via Parasitics		Parasitic Values for Different Bundle ARs of		
		1.2:1		5:1
N_{CNT}	45000	37522	100	18761
N_{total}	30015	25027	67	12513
R_{bundle} (Ω)	8.06	9.67	3628.1	3.22
L_{bundle} (fH)	4.61	5.52	2072.3	1.73
C_Q^{Bundle} (nF)	0.39	0.33	0.0008	0.03
C_E^{Bundle} (pF)	0.39	0.36	0.02	0.06
C_{Si} (fF)	10.96	10.96	10.96	0.07
G_{Si} (mho)	39.51	39.51	39.51	6.64

7.5 Impact of Polymer Liners on Delay

This section analyzes the delay with and without crosstalk for SiO_2 and different polymer liners. Crosstalk in coupled lines is broadly classified in two categories: (1) functional and (2) dynamic crosstalk. Under functional crosstalk category, the victim line experiences a voltage spike when the aggressor line switches. However, dynamic crosstalk is observed when the adjacent line (aggressor and victim) switches either in the same direction (in-phase) or in the opposite direction (out-phase).

Using an SWCNT bundled TSV with AR = 1.2:1, Figure 7.6a–c presents the out-phase delay, propagation delay, and in-phase delay, respectively [24]. It is observed that the out-phase delay is more compared to the in-phase and propagation delays. The reason behind this is the Miller capacitive effect that leads to almost doubling of coupling parasitics. Additionally, it is observed that the delay substantially reduces for higher number of SWCNTs in a bundle for a fixed via height and radius. It is due to the lower parasitic values that mainly depend on the number of conducting channels (N_{total}) of each SWCNT in the bundle. The higher N_{total} substantially reduces the overall resistance and inductance with a small increase in capacitance (Table 7.2). It results in reduced delay for bundled TSV having 45,000 SWCNTs.

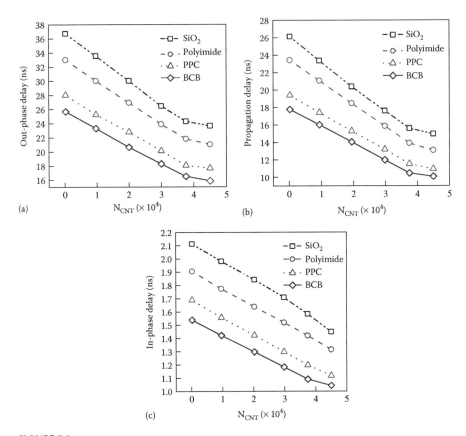

FIGURE 7.6
(a) Out-phase delay (OP), (b) propagation delay (PD), and (c) in-phase delay (IP) of SWCNT bundled TSVs having an AR = 1.2:1. (Kumari, A. et al., *Proceedings of the IEEE Electronics Components and Technology Conference (ECTC 2014)*, Orlando, FL, 2014.)

Using SiO_2 and different polymer liners, the propagation delay and crosstalk-induced delay of bundled TSV with AR = 1.2:1 ($N_{CNT} = 45000$) are compared with a TSV bundle having AR = 5:1 ($N_{CNT} = 18761$). Table 7.3 presents the percentage reduction in delay with and without crosstalk for different polymer liners. Compared to SiO_2 liner, the overall delay of bundled TSV with AR = 1.2:1 is reduced by 30.2%, 24.3%, and 10.9% for BCB, PPC, and polyimide, respectively; similarly, the overall reduction in delay for bundled TSV with AR = 5:1 is 43.8%, 36.2%, and 18%, respectively. The primary reason behind this reduced effect is the smaller quantitative values of oxide capacitances (Table 7.4). The oxide capacitances

TABLE 7.3

Percentage Reduction in Delay Using Polymer Liners with Respect to SiO_2 Liner

SWCNT Bundled TSV with	Delay	Reduction in Delay for Different Polymer Liners (%)		
		Polyimide	PPC	BCB
AR = 1.2:1 (N_{CNT} = 45,000)	OP	12.49	26.68	32.73
	PD	11.17	25.17	31.75
	IP	9.05	21.01	26.24
AR = 5:1 (N_{CNT} = 18,761)	OP	27.95	39.84	46.20
	PD	15.32	35.72	43.63
	IP	10.73	33.09	41.44

IP, in-phase delay; OP, out-phase delay; PD, propagation delay.

TABLE 7.4

Oxide Capacitances for Different Liner Materials

Liner Material	SWCNT Bundled TSV with	C_{OX} (fF)
SiO_2	AR = 1.2:1 (N_{CNT} = 45000)	10.936
	AR = 5:1 (N_{CNT} = 18761)	0.073
Polyimide	AR = 1.2:1 (N_{CNT} = 45000)	10.923
	AR = 5:1 (N_{CNT} = 18761)	0.062
PPC	AR = 1.2:1 (N_{CNT} = 45000)	10.917
	AR = 5:1 (N_{CNT} = 18761)	0.051
BCB	AR = 1.2:1 (N_{CNT} = 45000)	10.904
	AR = 5:1 (N_{CNT} = 18761)	0.04

are primarily dependent on the dielectric constant of the liner material between two TSVs. The lower dielectric constant of BCB substantially reduces the capacitance values, which results in lesser crosstalk-induced delay compared to SiO_2 and other polymer liners with relatively higher dielectric constant.

7.6 Summary

This chapter presented a comparative analysis of propagation delay and cross-talk-induced delay using SWCNT bundled TSV with SiO_2 and different polymer liners. It is observed that the crosstalk coupling is prominently influenced by the dielectric constant of the materials lying between the TSV and the Si substrate. Using BCB as a liner material, the oxide capacitance between two TSVs is reduced up to 34.48% compared to the conventional SiO_2 liner. It results in reduced propagation delay and crosstalk-induced delay for a higher bundle AR. For an SWCNT bundled TSV with a higher aspect ratio, the overall delay is reduced by 36.2% for PPC, whereas the reduction becomes 43.8% and 18% for BCB and polyimide, respectively, compared to the SiO_2 liner. The aforementioned results are obtained by considering only the effect of dielectric constant that demonstrates BCB as the most suitable liner material. However, PPC can be preferred over BCB and polyimide due to its advantage in obtaining the smooth surface, readily adjustable viscosity of precursors, good fluidity, and adhesion. These outstanding properties are the primary reasons that make PPC coating process complementary MOS compatible.

Multiple Choice Questions

1. The commonly used liner material in TSV is
   a. BCB
   b. SiO_2
   c. PPC
   d. Polyimide

2. BCB and PPC stand for
   a. Benzocyclobutene and polypropylene carbonate
   b. Benzocarbonbutene and polypropylene carbonate
   c. Benzocyclobutene and plastic-propylene carbonate
   d. None of these

3. The mechanical reliability problem such as die cracks or breakage may occur due to the
   a. Large differences in stress of the SiO_2 and the CNT bundles
   b. Small differences in the CTEs of the SiO_2 and the CNT bundles
   c. Large differences in the CTEs of the SiO_2 and the CNT bundles
   d. None of the above

4. Match the following:

Liner Material		**Relative Permittivity**	
A.	SiO_2	i.	3.5
B.	Polyimide	ii.	2.65
C.	PPC	iii.	3.9
D.	BCB	iv.	2.9

   a. A—i, B—ii, C—iii, D—iv
   b. A—ii, B—iii, C—i, D—iv
   c. A—iv, B—iii, C—ii, D—i
   d. A—iii, B—i, C—iv, D—ii

5. Polymers can be considered as suitable liner materials due to their
   a. Large and uniform thickness
   b. Low elastic modulus
   c. Both a and b
   d. None of the above

6. Polymer liner can act as a buffer layer between the silicon substrate and the metal due to its
   a. Large and uniform thickness
   b. Low elastic modulus
   c. Both a and b
   d. None of the above

7. The embedded TSVs on the Si substrate is surrounded by
   a. An isolation dielectric
   b. A depletion region
   c. An inversion layer
   d. None of the above

8. The thickness of this depletion region is primarily dependent on the
   a. applied voltage
   b. material properties
   c. interface charge density
   d. All of the above

9. The conduction mechanism in CNT is ballistic due to its
   a. Small radius in the range of nanometer
   b. Large current-carrying capability
   c. Long mpf in the range of micrometers
   d. None of these

10. The conductivity of the silicon substrate is
    a. 0.1 $(\Omega.\text{cm})^{-1}$
    b. 0.3 $(\Omega.\text{cm})^{-1}$
    c. 0.5 $(\Omega.\text{cm})^{-1}$
    d. 0.75 $(\Omega.\text{cm})^{-1}$

Short Questions

1. Name the different polymers that can be used as TSV liners.
2. What are the advantages of polymer liners over SiO_2 liners?
3. What is the role of the DRIE process during the fabrication of TSVs with polymer liners?
4. Discuss the steps of spin coating process during TSV fabrication.
5. What do you mean by the MOS effect of TSVs on the silicon substrate?
6. Define the contact and quantum resistance of CNT bundled TSVs and their typical values.
7. Why do we neglect the magnetic inductance with respect to the kinetic inductance of CNT-based TSVs?
8. Define the ballistic transport of CNT-based TSVs.
9. What are the roles of the quantum capacitance and the TSV capacitance for a CNT bundled TSV?
10. Define the functional and dynamic crosstalk effect between coupled TSV lines.
11. Why do the propagation delay and crosstalk-induced delay reduce for CNT bundled TSVs with higher aspect ratio?
12. Why do the researchers usually prefer PPC over BCB and polyimide as TSV liner materials?

Long Questions

1. Discuss in brief the different drawbacks of using SiO_2 liner for the CNT-based TSV fabrication process.
2. Describe the different fabrication steps for fabrication of TSVs with polymer liner.

3. Discuss the physical configuration and via parameters for the modeling of coupled CNT bundled signal–ground–signal TSVs.

4. Write down the model expressions with brief explanation for coupled CNT bundled TSVs.

5. Using a coupled CNT bundled TSVs, explain the impact of SiO_2, BCB, PPC, and polyimide liners on the propagation delay and crosstalk-induced delay.

References

1. Lu, J. Q. 2009. 3-D hyperintegration and packaging technologies for micro-nano systems. *Proceedings of the IEEE* 97(1):18–30.
2. Davis, W. R., Wilson, J., Mick, S., Xu, J., Hua, H, Mineo, C., Sule, A. M., Steer, M., and Franzon, P. D. 2010. Demystifying 3D ICs: The pros and cons of going vertical. *IEEE Design & Test of Computers* 22(6):498–510.
3. Jiang, H., Liu, B., Huang, Y., and Hwang, K. C. 2004. Thermal expansion of single wall carbon nanotubes: Special section on mechanics and mechanical properties of carbon nanotubes. *Journal of Engineering Materials and Technology* 126(3):265–270.
4. Li, H., Xu, C., Srivastava, N., and Banerjee, K. 2009. Carbon nanomaterials for next-generation interconnect and passives: Physics, status and prospects. *IEEE Transactions on Electron Devices* 56(9):1799–1821.
5. VanderPlas, G., Limaye, P., Loi, I. et al. 2010. Design issues and considerations for low-cost 3D TSV IC technology. *IEEE Journal of Solid State Circuits* 46(1):293–307.
6. Zhang, L., Gu, X., and J. Guo. 2011. Clinical observation on dental caries treatment using nanometer composite resin. In *Proceedings of the IEEE 2011 International Conference Human Health and Biomedical Engineering (HHBE 2011)*, August 19–22, pp. 668–670. Jilin, FL.
7. Hsin, Y. C., Chen, C. C., Lau, J. H. et al. 2011. Effects of etch rate on scallop of through-silicon vias (TSVs) in 200 mm and 300 mm wafers. In *Proceedings of the IEEE Electronic Components and Technology Conference (ECTC 2011)*, May 31–June 3, pp. 1130–1135. Lake Buena Vista, FL.
8. Ranganathan, N., Lee, D. Y., Youhe, L., Lo, G. Q., Prasad, K., and Pey, K. L. 2010. Influence of bosch etch process on electrical isolation of TSV structures. *IEEE Transactions on Components, Packaging and Manufacturing Technology* 46(1):293–307.
9. Bea, J., Lee, K., Fukushima, T., Tanaka, T., and Koyanagi, M. 2011. Evaluation of Cu diffusion from Cu through-silicon via (TSV) in three-dimensional LSI by transient capacitance measurement. *IEEE Electron Device Letters* 32(7):940–942.
10. Liu, X., Chen, Q., Dixit, P., Chatterjee, R., Tummala, R. R., and Sitaraman, S. K. 2009. Failure mechanisms and optimum design for electroplated copper through-silicon vias (TSV). In *Proceedings of the IEEE Electronic Components and Technology Conference (ECTC 2009)*, May 26–29, pp. 624–629. San diego, CA.

11. Selvanayagam, C. S., Lau, J. H., Zhang, X., Seah, S. K. W., Vaidyanathan, K., and Chai, T. C. 2009. Nonlinear thermal stress/strain analyses of copper filled TSV (through silicon via) and their flip-chip microbumps. *IEEE Transactions on Advanced Packaging* 32(4):720–728.

12. Shen, L. C., Chien, C. W., Cheng, H. C., and Lin, C. T. 2010. Development of three-dimensional chip stacking technology using a clamped through-silicon via interconnection. *Microelectronics Reliability* 50(4):489–497.

13. Ryu, S. K., Lu, K. H, Zhang, X., Im, J. H., Ho, P. S., and Huang, R. 2011. Impact of near-surface thermal stresses on interfacial reliability of through-silicon vias for 3-D interconnects. *IEEE Transactions on Device and Material Reliability* 11(1):35–43.

14. Thompson, S. E., Sun, G., Choi, Y. S., and Nishida, T. 2006. Uniaxial-process-induced strained-Si: Extending the CMOS roadmap. *IEEE Transactions on Electron Devices* 53(5):1010–1020.

15. Chen, Z., Song, X., and Liu, S. 2011. Thermo-mechanical characterization of copper filled and polymer filled TSVs considering nonlinear material behaviors. In *Proceedings IEEE Electronic Components and Technology Conference (ECTC 2009)*, May 26–29, pp. 1374–1380. San Diego, CA.

16. Burke, P. J. 2002. Lüttinger liquid theory as a model of the gigahertz electrical properties of carbon nanotubes. *IEEE Transactions on Nanotechnology* 1(3):129–144.

17. Tezcan, D. S., Duval, F., Philipsen, H., Luhn, O., Soussan, P., and Swinnen, B. 2009. Scalable through silicon via with polymer deep trench isolation for 3D wafer level packaging. In *Proceedings of the 59th Electronic Components and Technology Conference (ECTC 2009)*, May 26–29, pp. 1159–1164. San Diego, CA.

18. Duval, F. F. C., Okoro, C., Civale, Y., Soussan, P., and Beyne, E. 2011. Polymer filling of silicon trenches for 3-D through silicon vias applications. *IEEE Transactions on Components, Packaging and Manufacturing Technology* 1(6):825–832.

19. Srivastava, A., Xu, Y., and Sharma, A. K. 2010. Carbon nanotubes for next generation very large scale interconnects. *Journal of Nanophotonics* 4(041690):1–26.

20. Sarto, M. S., and Tamburrano, A. 2010. Single-conductor transmission-line model of multiwall carbon nanotubes. *IEEE Transactions on Nanotechnology* 9(1):82–92.

21. Naeemi, A., and Meindl, J. D. 2008. Performance modeling for single- and multi-wall carbon nanotubes as signal and power interconnects in gigascale systems. *IEEE Transactions on Electron Devices* 55(10):2574–2582.

22. Zhao, W. S., Yin, W. Y., and Guo, Y. X. 2012. Electromagnetic compatibility-oriented study on through silicon single-walled carbon nanotube bundle via (TS-SWCNTBV) arrays. *IEEE Transactions on Electromagnetic Compatibility* 54(1):149–157.

23. Kannan, S., Gupta, A., Kim, B. C., Mohammed, F., and Ahn, B. 2010. Analysis of carbon naotube based through silicon vias. In *Proceedings of the IEEE 60th Electronic Components and Technology Conference (ECTC 2010)*, June 1–4, pp. 51–57. Las Vegas, NV.

24. Kumari, A., Majumder, M. K., Kaushik, B. K., and Manhas, S. K. 2014. Effect of polymer liners in CNT based through silicon vias. In *Proceedings of the IEEE Electronics Components and Technology Conference (ECTC 2014)*, May 27–30, pp. 1921–1925. Orlando, FL.

8

Modeling of Through Silicon Vias Using Finite-Difference Time-Domain Technique

8.1 Introduction to Finite-Difference Time-Domain Technique

Finite-difference time-domain (FDTD) is a popular computational electrodynamics modeling technique. Initially, this method was implemented to solve Maxwell's equations. In this technique, the time-dependent partial differentials are discretized using central-difference approximations to the space and time partial derivatives. The resulting finite-difference equations are solved in either software or hardware in a leapfrog manner. For the first time, this method was implemented by Yee et al. [1] for Maxwell's equations solutions. In this method, FDTD is implemented on the transmission line equation in which relative parameters are line voltage (V) and current (I) similar to electric field (E) and magnetic field (H) in Maxwell's equations.

To implement an FDTD solution of through silicon vias (TSVs), a computational domain must be established. The computational domain is simply the physical region over which the model will be performed. In the modeling of TSVs, the transmission line is the computational domain, in which V and I are the relative parameters. The V and I fields are determined at every point in space within that computational domain. The transmission line parasitic such as resistance (R), capacitance (C), conductance (G), and inductance (L) of each cell within the computational domain must be specified. For coupled TSV line configuration, the capacitive coupling factor (C_c) should also be added in the computational domain.

Once the computational domain is established, the source and load terminal conditions must be specified. The source can be either linear or nonlinear voltage or current source, and the load is a capacitor in case of TSVs. After specifying the load and source terminals, the boundary conditions are used at the interfaces of computational domain to match the FDTD solution.

8.1.1 Working with the FDTD Method

When transmission TSV line equations are examined, it can be noticed that the change in the voltage (V) in time (the time derivative) is dependent on the change in the current (I) across space (the curl). At any point in space, the updated value of V in time is dependent on the stored value of V and the numerical curl of the local distribution of I in space. The I is time stepped in a similar manner. At any point in space, the updated value of I in time is dependent on the stored value of I and the numerical curl of the local distribution of V in space. Iterating I and V updates results in a explicit marching-in-time process, in which sampled data analogs of the continuous electromagnetic signal under consideration propagate in a numerical grid stored in the computer memory.

8.1.2 Stability Criterion for FDTD

The numerical stability of the FDTD method is determined by the Courant–Friedrichs–Lewy (CFL) condition, which simply requires that a signal cannot be allowed to travel more than one space size during one time step. The CFL condition is a necessary condition for convergence while solving certain partial differential equations numerically by the method of finite differences. It arises when explicit time-matching schemes are used for the numerical solution [2].

For stability of the FDTD solution, the position and time discretization must satisfy the Courant condition. Numerically, the time and space discretization should follow the following equation [2]:

$$\Delta t \le \frac{\Delta z}{v} \tag{8.1}$$

where Δt, Δz and v represent time discretization, space discretization and signal velocity in line, respectively.

The Courant stability condition provides that for stability of the solution the time step must be no greater than the propagation time over each cell.

8.1.3 Central Difference Approximation

Consider a function of one variable $f(t)$. Expanding this in a Taylor series in a neighborhood of a desired point t_0 gives

$$f(t_0 + \Delta t) = f(t_0) + \Delta t f'(t_0) + \frac{\Delta t^2}{2!} f''(t_0) + \frac{\Delta t^3}{3!} f'''(t_0) + \cdots \tag{8.2}$$

where the primes denote the various derivatives with respect to of the function. Solving this for the first derivative gives

$$f'(t_0) = \frac{f(t_0 + \Delta t) - f(t_0)}{\Delta t} - \frac{\Delta t}{2!} f''(t_0) - \frac{\Delta t^2}{3!} f'''(t_0) + \cdots \tag{8.3}$$

Thus, the first-order derivative can be approximated as

$$f'(t_0) = \frac{f(t_0 + \Delta t) - f(t_0)}{\Delta t} + \theta(\Delta t) \tag{8.4}$$

where $\theta(\Delta t)$ denotes that the error in truncating the series is on the order of Δt. Therefore, the first-order derivative may be approximated with the forward difference:

$$f'(t_0) \cong \frac{f(t_0 + \Delta t) - f(t_0)}{\Delta t} \tag{8.5}$$

This amounts to approximating the derivative of $f(t)$ as with its region of the desired point. Using Taylor's series expansion,

$$f(t_0 - \Delta t) = f(t_0) - \Delta t f'(t_0) + \frac{\Delta t^2}{2!} f''(t_0) - \frac{\Delta t^3}{3!} f'''(t_0) + \cdots \tag{8.6}$$

which can be approximated by

$$f'(t_0) \cong \frac{f(t_0) - f(t_0 - \Delta t)}{\Delta t} \tag{8.7}$$

This gives the backward approximate equation for the derivative.

Other approximations known as central differences can be found by subtracting Equation 8.2 from Equation 8.6 to yield the central difference approximation for the first derivative:

$$f'(t_0) \cong \frac{f(t_0 + \Delta t) - f(t_0 - \Delta t)}{2\Delta t} \tag{8.8}$$

with a truncation error of Δt^2. Similarly, the central difference approximation for the second derivative is obtained by adding Equations 8.2 and 8.6 to yield

$$f''(t_0) \cong \frac{f(t_0 + \Delta t) - 2f(t_0) + f(t_0 - \Delta t)}{\Delta t^2} \tag{8.9}$$

with a truncation error on the order of Δt^2. Due to the second-order error truncation in the central difference approximation, FDTD is termed as FDTD solution with the second-order accuracy.

8.2 FDTD Model

This section presents the formalization of an FDTD method that is intended for estimation of voltages and currents on TSVs. The TSVs are modeled as an *RLCG* transmission line [3].

8.2.1 TSV Line Equation

In general, TSV behaves as a waveguide that can be analyzed using Maxwell's equations. TSV can also be analyzed using the transmission line equations, if it is assumed that the waves on the line propagate in the transverse electromagnetic (TEM) mode [1]. TSV models based on transmission lines with distributed *RLGC* segments are thus normally used in the analysis.

Figure 8.1 shows the *RLGC* distributed transmission line, where *R, L, G,* are *C* are per-unit-length resistance, inductance, conductance, and capacitance of the transmission line of length *l*, respectively. Consider the point *z* along the transmission line at time *t*. The following set of equations holds:

$$\frac{d}{dz}V(z,t) = -RI(z,t) - L\frac{d}{dt}I(z,t) \tag{8.10}$$

$$\frac{d}{dz}I(z,t) = -GV(z,t) - C\frac{d}{dt}V(z,t) \tag{8.11}$$

8.2.2 Discretization in Space and Time

To implement the FDTD solution for Equation 8.10, first the transmission line must be divided into *Nz* section, each of length Δz as shown in Figure 8.2. Similarly, the total solution time is divided into *Nt* segments of length Δt [3]. Discretization of voltage and current and relation with time and space are

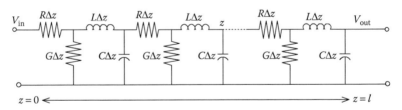

FIGURE 8.1
Distributed *RLGC* TSV line.

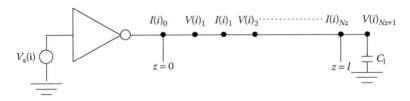

FIGURE 8.2
Illustration of the discretization of the *i*th TSV line for the implementation of FDTD.

shown in Figure 8.3. In order to ensure stability of the discretization and to ensure the second-order accuracy, the $Nz + 1$ voltage points, $V_1, V_2,..., V_{Nz}$, and the Nz current points, $I_1, I_2,..., I_{Nz}$, are need to be interlaced. Each voltage and adjacent current solution point is separated by $\Delta z/2$. The time points are also interlaced, and each voltage time point and adjacent current time point are separated by $\Delta t/2$, as also shown in Figure 8.3. Now using central differential approximation on Equation 8.10,

$$\frac{V_{K+1}^{n+1} - V_K^{n+1}}{\Delta z} + L\frac{I_K^{n+3/2} - I_K^{n+1/2}}{\Delta t} + R\frac{I_K^{n+3/2} + I_K^{n+1/2}}{2} = 0 \tag{8.12}$$

for $K = 1, 2, 3,..., Nz$, and similarly applying central difference approximation on Equation 8.11 becomes

$$\frac{I_K^{n+1/2} - I_{K-1}^{n+1/2}}{\Delta z} + C\frac{V_K^{n+1} - V_K^n}{\Delta t} + G\frac{V_K^{n+1} + V_K^n}{2} = 0 \tag{8.13}$$

for $K = 2, 3, 4,..., Nz$, where the voltage and current are denoted as

$$V_K^n = V\left[(K-1)\Delta z, n\Delta t\right] \tag{8.14}$$

$$I_K^n = I\left[(K-1/2)\Delta z, n\Delta t\right] \tag{8.15}$$

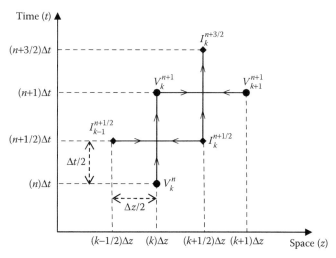

FIGURE 8.3
Interlacing the current and voltage solutions in space and time for the FDTD analysis.

By rearranging Equation 8.12, the recursive relation of current at an interior point of TSV can be derived as

$$\frac{V_{K+1}^{n+1} - V_K^{n+1}}{\Delta z} + L\frac{I_K^{n+3/2} - I_K^{n+1/2}}{\Delta t} + R\frac{I_K^{n+3/2} + I_K^{n+1/2}}{2} = 0$$

$$\frac{1}{\Delta z}\left[V_{k+1}^{n+1} - V_K^{n+1}\right] + \left[\frac{L}{\Delta t} + \frac{R}{2}\right]I_k^{n+3/2} + \left[\frac{R}{2} - \frac{L}{\Delta t}\right]I_k^{n+1/2} = 0 \qquad (8.16)$$

$$\left[\frac{L}{\Delta t} + \frac{R}{2}\right]I_k^{n+3/2} = \frac{1}{\Delta z}\left[V_k^{n+1} - V_{k+1}^{n+1}\right] + \left[\frac{L}{\Delta t} - \frac{R}{2}\right]I_k^{n+1/2} \qquad (8.17)$$

$$I_k^{n+3/2} = EFI_k^{n+1/2} + \frac{E}{\Delta z}\left(V_k^{n+1} - V_{k+1}^{n+1}\right) \qquad (8.18)$$

for $K = 2, 3, 4, \ldots, Nz$
where:

$$E = \left(\frac{L}{\Delta t} + \frac{R}{2}\right)^{-1} \text{ and } F = \left(\frac{L}{\Delta t} - \frac{R}{2}\right) \qquad (8.19)$$

By rearranging Equation 8.13, the recursive relation of voltage at an interior point of TSV can be derived as

$$\frac{I_K^{n+1/2} - I_{K-1}^{n+1/2}}{\Delta z} + C\frac{V_K^{n+1} - V_K^n}{\Delta t} + G\frac{V_K^{n+1} + V_K^n}{2} = 0$$

$$\left[\frac{C}{\Delta t} + \frac{G}{2}\right]V_k^{n+1} + \left[\frac{G}{2} - \frac{C}{\Delta t}\right]V_k^n + \frac{1}{\Delta z}\left[I_k^{n+(1/2)} - I_{k-1}^{n+(1/2)}\right] = 0 \qquad (8.20)$$

$$\left[\frac{C}{\Delta t} + \frac{G}{2}\right]V_k^{n+1} = \left[\frac{C}{\Delta t} - \frac{G}{2}\right]V_k^n + \frac{1}{\Delta z}\left[I_{k-1}^{n+(1/2)} - I_k^{n+(1/2)}\right] \qquad (8.21)$$

$$V_K^{n+1} = \left[\frac{C}{\Delta t} - \frac{G}{2}\right]\left[\frac{C}{\Delta t} - \frac{G}{2}\right]^{-1}V_k^n + \frac{1}{\Delta z}\left[\frac{C}{\Delta t} - \frac{G}{2}\right]^{-1}\left[I_{k-1}^{n+(1/2)} - I_k^{n+(1/2)}\right] \qquad (8.22)$$

$$V_k^{n+1} = ABV_k^n + \frac{A}{\Delta z}\left(I_{k-1}^{n+1/2} - I_k^{n+1/2}\right) \qquad (8.23)$$

for $K = 2, 3, \ldots, Nz$
where

$$A = \left(\frac{C}{\Delta t} + \frac{G}{2}\right)^{-1} \text{ and } B = \left(\frac{C}{\Delta t} - \frac{G}{2}\right) \qquad (8.24)$$

Equations 8.18 and 8.23 are used to estimate the voltage and current at different points of line at different times for signal transmission line. This can be extended for multitransmission lines as well. For N transmission lines, the voltage and current can be replaced by the vector of order of $1 \times N$ as

$$V = [V(1)V(2)V(3)...V(N)]^{T} \tag{8.25}$$

$$I = [I(1)I(2)I(3)...I(N)]^{T} \tag{8.26}$$

where $V(1)$, $V(2)$,..., $V(N)$ are the voltage points for lines 1, 2,... N, respectively, similarly apply for current points. Line parameters such as R, L, G, and C and all other derived parameters such as A, B, E, and F are replaced by $N \times N$ matrix for N transmission lines.

8.2.3 Leapfrog Time Stepping

As discussed in Section 8.2.2, the current is updated at $t = (n+1/2)\Delta t$ using the previous current value at $t = (n-1/2)\Delta t$ and the voltage at $t = n\Delta t$. However, the voltage is updated at $t = (n+1)\Delta t$ using the previous value at $t = n\Delta t$ and the current at $t = (n+1/2)\Delta t$. Figure 8.4 shows the timeline when I and V are updated.

To remain numerically stable, the time step must satisfy the Courant stability condition [2]:

$$\Delta t \leq \frac{1}{C\Delta z} \tag{8.27}$$

To obtain good spatial resolution, the cell size should be less than a tenth of the shortest wavelength:

$$\Delta z < \frac{\lambda_{min}}{10} \tag{8.28}$$

8.2.4 Incorporation of Boundary Conditions

This section considers the incorporation of boundary conditions in FDTD solution. The essential problem in incorporating the terminal conditions is that the FDTD voltages and currents at each end of the line, V_1, I_1, and V_{Nz+1},

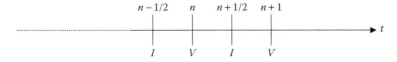

FIGURE 8.4
Timeline showing when V and I fields are updated on a TSV line.

I_{Nz}, are not collocated in space or time, whereas the terminal conditions relate the voltage and current at the same position and at the same time [4]. This section can be divided into two parts: boundary matching at the source and line interface and boundary matching at the load and line interface.

8.2.4.1 Boundary Matching at the Source End

Let the source current be I_s at the source terminal ($z = 0$) as shown in Figure 8.5. To match the source current at the boundary, the current $I_0^{n+1/2}$ can be discretized by averaging the source currents I_s

$$I_0^{n+1/2} = \frac{I_s^{n+1} + I_s^n}{2} \tag{8.29}$$

The discretization of currents at source and load terminals is shown in Figure 8.6. Now using Equations 8.13 and 8.29, the recursive relation for the voltage and current at the source and line interface can be written as

$$\frac{1}{\Delta z/2}\left(I_1^{n+1/2} - \frac{I_s^{n+1} + I_s^n}{2}\right) + \frac{1}{\Delta t}C\left(V_1^{n+1} - V_1^n\right) + \frac{G}{2}\left(V_1^{n+1} + V_1^n\right) = 0 \tag{8.30}$$

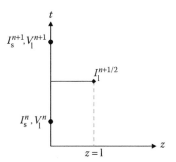

FIGURE 8.5
Discretizing the terminal voltages and currents of the TSV line at the source end in order to incorporate the terminal constraints.

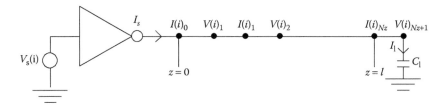

FIGURE 8.6
Discretizing of TSV line with source and load currents.

8.2.4.2 Boundary Matching at the Load End

The current at the load terminal is shown in Figure 8.7. The transmission line equation is discretized at the load by averaging the load current I_1. In order to obtain a value that is located in time at the same point at $I_{Nz+1}^{n+1/2}$,

$$I_{Nz+1}^{n+1/2} = \frac{I_1^{n+1} + I_1^n}{2} \tag{8.31}$$

Now using Equations 8.13 and 8.31, the recursive relation between the voltage and the current at the load end and line interface can be written as

$$\frac{1}{\Delta z/2}\left(\frac{I_1^{n+1}+I_1^n}{2}-I_{Nz}^{n+1/2}\right)+\frac{1}{\Delta t}C\left(V_{Nz+1}^{n+1}-V_{Nz+1}^n\right)+\frac{G}{2}\left(V_{Nz+1}^{n+1}+V_{Nz+1}^n\right)=0 \tag{8.32}$$

8.2.5 FDTD Model for TSV Terminated by a Resistive Load

Let TSV be terminated by a resistor at the load end. Using Figure 8.8, the relation between the load current I_1 and the load voltage V_1 can be written as

$$V_{Nz+1} = V_1 + I_1 R_1 \tag{8.33}$$

Discretizing the above equation into the time domain,

$$V_{Nz+1}^{n+1} = V_1^{n+1} + I_1^{n+1} R_1 \tag{8.34}$$

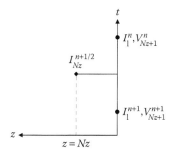

FIGURE 8.7
Discretizing the terminal voltages and currents of the line at the load end in order to incorporate the terminal constraints.

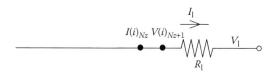

FIGURE 8.8
TSV terminated by resistive load.

On rearranging the above equation,

$$I_1^{n+1} = \frac{V_{Nz+1}^{n+1} - V_1^{n+1}}{R_1} \tag{8.35}$$

$$I_1^n = \frac{V_{Nz+1}^n - V_1^n}{R_1} \tag{8.36}$$

Now using the above equation and the relation derived in Equation 8.32, the recursive relation for the voltage at resistive load can be derived as

$$\frac{1}{\Delta z/2}\left(\frac{I_1^{n+1}+I_1^n}{2} - I_{Nz}^{n+1/2}\right) + \frac{1}{\Delta t}C\left(V_{Nz+1}^{n+1} - V_{Nz+1}^n\right) + \frac{G}{2}\left(V_{Nz+1}^{n+1} + V_{Nz+1}^n\right) = 0$$

$$\frac{2}{\Delta z}\left(\frac{V_{Nz+1}^{n+1} - V_1^{n+1} + V_{Nz+1}^n - V_1^n}{2R_1} - I_{Nz}^{n+1/2}\right) + \frac{C}{\Delta t}\left(V_{Nz+1}^{n+1} - V_{Nz+1}^n\right) + \frac{G}{2}\left(V_{Nz+1}^{n+1} + V_{Nz+1}^n\right) = 0 \tag{8.37}$$

$$\left(\frac{1}{R_1\Delta z} + \frac{C}{\Delta t} + \frac{G}{2}\right)V_{Nz+1}^{n+1} - \left(\frac{C}{\Delta t} - \frac{G}{2} - \frac{1}{R_1\Delta z}\right)V_{Nz+1}^n - \frac{2}{\Delta z}\left(I_{Nz}^{n+1/2} + \frac{V_1^{n+1}+V_1^n}{2R_1}\right) = 0 \tag{8.38}$$

$$\left(\frac{1}{R_1\Delta z} + \frac{C}{\Delta t} + \frac{G}{2}\right)V_{Nz+1}^{n+1} = \left(\frac{C}{\Delta t} - \frac{G}{2} - \frac{1}{R_1\Delta z}\right)V_{Nz+1}^n + \frac{2}{\Delta z}\left(I_{Nz}^{n+1/2} + \frac{V_1^{n+1}+V_1^n}{2R_1}\right) \tag{8.39}$$

$$V_{Nz+1}^{n+1} = K_1 K_2 V_{Nz}^n + \frac{2K_1}{\Delta z}\left(I_{Nz}^{n+1/2} + \frac{V_1^n - V_1^{n+1}}{2R_1}\right) \tag{8.40}$$

where:

$$K_1 = \left(\frac{1}{R_1\Delta z} + \frac{C}{\Delta t} + \frac{G}{2}\right)^{-1} \text{ and } K_2 = \left(\frac{C}{\Delta t} - \frac{G}{2} - \frac{1}{R_1\Delta z}\right) \tag{8.41}$$

8.2.6 FDTD Model for TSV Terminated by a Capacitive Load

In the second case, the TSV is terminated by the capacitor at the far end. Using Figure 8.9, the relation between the load current I_1 and the load voltage V_{Nz+1} can be written as

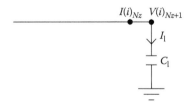

FIGURE 8.9
TSV terminated by capacitive load.

$$I_1 = C_1 \frac{d}{dt} V_{Nz+1} \tag{8.42}$$

Discretizing the above equation in the time domain,

$$I_1^{n+1} = C_1 \frac{\left(V_{Nz+1}^{n+1} - V_{Nz+1}^{n}\right)}{\Delta t} \tag{8.43}$$

Now, using Equations 8.32 and 8.43, the interface equation for the voltage at the capacitive load end is derived as

$$\frac{1}{\Delta z/2}\left(\frac{I_1^{n+1}+I_1^{n}}{2}-I_{Nz}^{n+1/2}\right)+\frac{1}{\Delta t}C\left(V_{Nz+1}^{n+1}-V_{Nz+1}^{n}\right)+\frac{G}{2}\left(V_{Nz+1}^{n+1}+V_{Nz+1}^{n}\right)=0$$

$$\frac{2}{\Delta z}\left[\frac{C_1\left(V_{Nz+1}^{n+1}-V_{Nz+1}^{n}\right)}{2\Delta t}+\frac{I_1^{n}}{2}-I_{Nz}^{n+1/2}\right]+\frac{1}{\Delta t}C\left(V_{Nz+1}^{n+1}-V_{Nz+1}^{n}\right)$$

$$+\frac{G}{2}\left(V_{Nz+1}^{n+1}+V_{Nz+1}^{n}\right)=0 \tag{8.44}$$

$$\left(\frac{C_1}{\Delta t\Delta z}+\frac{C}{\Delta t}+\frac{G}{2}\right)V_{Nz+1}^{n+1}-\left(\frac{C_1}{\Delta z\Delta t}+\frac{C}{\Delta t}-\frac{G}{2}\right)V_{Nz+1}^{n}-\frac{2}{\Delta z}\left(I_{Nz}^{n+1/2}-\frac{I_1^{n}}{2}\right)=0 \tag{8.45}$$

$$\left(\frac{C_1}{\Delta t\Delta z}+\frac{C}{\Delta t}+\frac{G}{2}\right)V_{Nz+1}^{n+1}=\left(\frac{C_1}{\Delta z\Delta t}+\frac{C}{\Delta t}-\frac{G}{2}\right)V_{Nz+1}^{n}+\frac{2}{\Delta z}\left(I_{Nz}^{n+1/2}-\frac{I_1^{n}}{2}\right) \tag{8.46}$$

$$V_{Nz+1}^{n+1}=K_1K_2V_{Nz+1}^{n}+\frac{2K_1}{\Delta z}\left(I_{Nz}^{n+1/2}-\frac{I_1^{n}}{2}\right) \tag{8.47}$$

where:

$$K_1=\left(\frac{C_1}{\Delta t\Delta z}+\frac{C}{\Delta t}+\frac{G}{2}\right)^{-1} \text{ and } K_2=\left(\frac{C_1}{\Delta z\Delta t}+\frac{C}{\Delta t}-\frac{G}{2}\right) \tag{8.48}$$

8.2.7 FDTD Model for TSV Driven by a Resistive Driver

Complementary metal–oxide–semiconductor (CMOS) driver is usually represented by a linear circuit, which uses the equivalent P-channel metal–oxide–semiconductor (PMOS) resistance for falling ramp input or the equivalent N-channel metal–oxide–semiconductor (NMOS) resistance for rising ramp input. In CMOS gate-driven TSV, for simplification the CMOS driver is replaced by its equivalent on resistance in series with power supply. The equivalent MOS resistance R_{eq} can be approximated by averaging the values of the resistance at the midpoint (where MOS transistor is in saturation state) during switching, which is in the form of [5]

$$R_{eq} = \frac{3}{4}\frac{V_{dd}}{I_{sat}}\left(1 - \frac{7}{9}\lambda V_{dd}\right) \tag{8.49}$$

where:
V_{dd} is the power supply voltage
I_{sat} is the drain saturation current
λ is the channel length modulation

The termination circuit is shown in Figure 8.10, which at the source end of TSV line consists of a resistor and a voltage source.
The voltage, V_1, can be expressed as

$$V_1 = V_s - I_s R_{eq} \tag{8.50}$$

On rearranging the above equation, the source current, I_s, can be expressed as

$$I_s = \frac{V_s - V_1}{R_{eq}} \tag{8.51}$$

To implement the FDTD for the above equation, it should be discretized in the time domain as

$$I_s^{n+1} = \frac{V_s^{n+1} - V_1^{n+1}}{R_{eq}} \tag{8.52}$$

$$I_s^n = \frac{V_s^n - V_1^n}{R_{eq}} \tag{8.53}$$

Using Equations 8.30, 8.52, and 8.53, recursion relations at the source interface for V_1 can be given as

$$\frac{1}{\Delta z/2}\left(I_1^{n+1/2} - \frac{I_s^{n+1} + I_s^n}{2}\right) + \frac{1}{\Delta t}C\left(V_1^{n+1} - V_1^n\right) + \frac{G}{2}\left(V_1^{n+1} + V_1^n\right) = 0$$

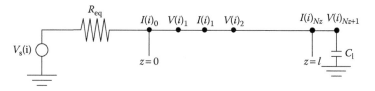

FIGURE 8.10
TSV driven by resistive driver.

$$\frac{2}{\Delta z}\left[I_1^{n+1/2} + \frac{V_s^{n+1} + V_s^n - \left(V_1^{n+1} + V_1^n\right)}{2R_{eq}}\right] + \frac{C}{\Delta t}\left(V_1^{n+1} - V_1^n\right) + \frac{G}{2}\left(V_1^{n+1} + V_1^n\right) = 0 \quad (8.54)$$

$$\left(\frac{-1}{\Delta z R_{eq}} + \frac{C}{\Delta t} + \frac{G}{2}\right)V_1^{n+1} - \left(\frac{1}{\Delta z R_{eq}} + \frac{C}{\Delta t} - \frac{G}{2}\right)V_1^n + \left(\frac{2}{\Delta z}\right)I_1^{n+1/2}$$

$$+ \frac{1}{\Delta z R_{eq}}\left(V_s^{n+1} + V_s^n\right) = 0 \quad (8.55)$$

$$\left(\frac{-1}{\Delta z R_{eq}} + \frac{C}{\Delta t} + \frac{G}{2}\right)V_1^{n+1} = \left(\frac{1}{\Delta z R_{eq}} + \frac{C}{\Delta t} - \frac{G}{2}\right)V_1^n - \left(\frac{2}{\Delta z}\right)I_1^{n+1/2}$$

$$- \frac{1}{\Delta z R_{eq}}\left(V_s^{n+1} + V_s^n\right) \quad (8.56)$$

$$V_1^{n+1} = K_1 K_2 V_1^n - \frac{K_1}{\Delta z R_{eq}}(V_s^n + V_s^{n+1}) - \frac{2K_1}{\Delta z}R_{eq}I_1^{n+1/2} \quad (8.57)$$

$$K_1 = \left(\frac{-1}{\Delta z R_{eq}} + \frac{C}{\Delta t} + \frac{G}{2}\right)^{-1} \text{ and } K_2 = \left(\frac{1}{\Delta z R_{eq}} + \frac{C}{\Delta t} - \frac{G}{2}\right) \quad (8.58)$$

8.2.8 FDTD Model for CMOS Gate-Driven TSV

In Section 8.2.7, the CMOS driver is simplified as a linear circuit in which a constant resistance is used to approximate the nonlinear and time-varying MOS resistance. FDTD is implemented for linear behavior of driver, whereas this section presents an FDTD method for transient analysis of lossy transmission lines in the presence of the nonlinear behavior of CMOS gates.

Signal integrity is strongly affected by both the transmission line behavior of TSVs and the nonlinear behavior of CMOS drivers. Therefore, accurate prediction of time delay and crosstalk noise should combine the effect of transmission line that propagates quasi-TEM with precise transistor-level modeling of CMOS gate. The nonlinear behavior of CMOS gates is represented by alpha-power law model [6], with the drain current described by the piecewise linear function of the drain voltage and discretized in the time domain for FDTD implementation.

Figure 8.11 shows M coupled TSVs driven by CMOS gates. The input signal is $N \times 1$ vector of voltage, denoted as $V_s = \left[V_s(1)V_s(2)V_s(3)\dots V_s(N)\right]^{-1}$. The modified alpha-power law model [6] is used for modeling the CMOS gates. Figure 8.12 shows the macro model used for FDTD implementation.

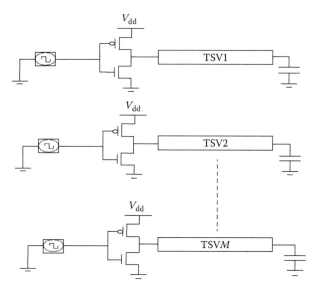

FIGURE 8.11
M coupled TSVs driven by CMOS gates.

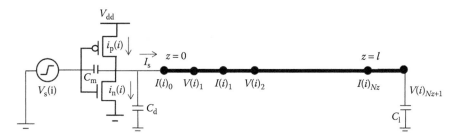

FIGURE 8.12
Macro model for CMOS-gate-driven TSV.

PMOS and NMOS currents are represented by the modified alpha-power law MOS model and discretized in the time domain, shown in the following two equations, respectively:

$$
i_p = \begin{cases}
0 & V_s \geq V_{dd} - \left| V_{tp} \right| & \text{(off)} \\[2mm]
K_{lp}\left(V_{dd} - V_s - \left| V_{tp} \right|\right)^{\alpha_p/2}\left(V_{dd} - V_0\right) & V_0 > V_{dd} - V_{DSATp} & \text{(lin)} \\[2mm]
K_{Sp}\left(V_{dd} - V_s - \left| V_{tp} \right|\right)^{\alpha_p}\left(1 + \sigma_p\left(V_{dd} - V_0\right)\right) & V_0 \leq V_{dd} - V_{DSATp} & \text{(sat)}
\end{cases}
\quad (8.59)
$$

$$i_n = \begin{cases} 0 & V_s \le V_{tn} \quad \text{(off)} \\ K_{ln}\left(V_s - V_{tn}\right)^{\alpha_n/2} V_0 & V_0 < V_{Dsatn} \quad \text{(lin)} \\ K_{sn}\left(V_s - V_{tn}\right)^{\alpha_n}\left(1 + \sigma_n V_0\right) & V_0 \ge V_{Dsatn} \quad \text{(sat)} \end{cases} \tag{8.60}$$

where:

σ represents the drain conductance

V_{Dsat} is the saturation voltage, which can be expressed as

$$V_{Dsat_p}^n = \frac{K_{sp}}{K_{lp}}\left(V_{dd} - V_s^n - \left|V_{tp}\right|\right)^{\alpha_p/2} \tag{8.61}$$

$$V_{Dsat_n}^n = \frac{K_{sn}}{K_{ln}}\left(V_s^n - V_{tn}\right)^{\alpha_n/2} \tag{8.62}$$

Applying Kirchoff's current law at the source and TSV interface, according to Figure 8.12, results in

$$I_s = C_m\left[\frac{d(V_s - V_1)}{dt}\right] + i_p - i_n - C_d\left(\frac{dV_1}{dt}\right) \tag{8.63}$$

where C_m and C_d are the drain-to-gate coupling capacitance and the drain diffusion capacitance of CMOS driver, respectively. The source current, I_s, in discretized form can be written as

$$I_s^{n+1} = C_m\frac{V_s^{n+1} - V_s^n}{\Delta t} + i_p^{n+1} - i_n^{n+1} - (C_m + C_d)\left(\frac{V_1^{n+1} - V_1^n}{\Delta t}\right) \tag{8.64}$$

Now solving Equations 8.64 and 8.30, the recursive relation for the voltage at the CMOS gate and TSV interface can be given as

$$\frac{1}{\Delta z/2}\left(I_1^{n+1/2} - \frac{I_s^{n+1} + I_s^n}{2}\right) + \frac{1}{\Delta t}C\left(V_1^{n+1} - V_1^n\right) + \frac{G}{2}\left(V_1^{n+1} + V_1^n\right) = 0$$

$$\frac{2}{\Delta z}\left[I_1^{n+1/2} - \frac{C_m\dfrac{V_s^{n+1} - V_s^n}{\Delta t} + i_p^{n+1} - i_n^{n+1} - (C_m + C_d)\dfrac{V_1^{n+1} - V_1^n}{\Delta t}}{2} + \frac{I_s^n}{2}\right]$$

$$= \frac{-1}{\Delta t}C\left(V_1^{n+1} - V_1^n\right) + \frac{-G}{2}\left(V_1^{n+1} + V_1^n\right) \tag{8.65}$$

$$V_1^{n+1} = K_1 V_1^n - K_2 I_1^{n+1/2} + \frac{K_2}{2}\left(i_p^{n+1} - i_n^{n+1}\right) + \frac{K_2}{2}\left[I_s^n + \frac{C_m}{\Delta t}\left(V_s^{n+1} - V_s^n\right)\right] \quad (8.66)$$

where:

$$K_1 = \left[\Delta z \Delta t U + A(C_m + C_1)\right]^{-1}\left[\Delta z \Delta t AB + A(C_m + C_1)\right]$$

$$K_2 = 2A\Delta t\left[\Delta z \Delta t U + A(C_m + C_1)\right]^{-1}$$

$$A = \left(\frac{C}{\Delta t} + \frac{G}{2}\right)^{-1}, \ B = \left(\frac{C}{\Delta t} - \frac{G}{2}\right) \quad (8.67)$$

8.2.9 FDTD Model for Coupled Transmission Line

In this section, two distributed coupled TSV lines are considered. Subscripts 1 and 2 are used for TSV1 and TSV2, respectively. C_c and M represent the coupling capacitance per unit height and the mutual inductance per unit height between the lines [7]. At any point z along the line, voltage and current waveforms on line 1 and line 2 satisfy the following set of differential equations (Figure 8.13):

$$\frac{\partial}{\partial z}V_1(z,t) + L_1\frac{\partial}{\partial t}I_1(z,t) + M\frac{\partial}{\partial t}I_2(z,t) + R_1I_1(z,t) = 0 \quad (8.68)$$

$$\frac{\partial}{\partial z}V_2(z,t) + L_2\frac{\partial}{\partial t}I_2(z,t) + M\frac{\partial}{\partial t}I_1(z,t) + R_2I_2(z,t) = 0 \quad (8.69)$$

$$\frac{\partial}{\partial z}I_1(z,t) + (C_1 + C_c)\frac{\partial}{\partial t}V_1(z,t) - C_c\frac{\partial}{\partial t}V_2(z,t) + G_1V_1(z,t) = 0 \quad (8.70)$$

$$\frac{\partial}{\partial z}I_2(z,t) + (C_2 + C_c)\frac{\partial}{\partial t}V_2(z,t) - C_c\frac{\partial}{\partial t}V_1(z,t) + G_2V_2(z,t) = 0 \quad (8.71)$$

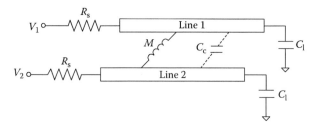

FIGURE 8.13
Coupled TSV transmission line model.

which can be written in matrix form as

$$\frac{\partial}{\partial z}V(z,t)+L\frac{\partial}{\partial t}I(z,t)+RI(z,t)=0 \tag{8.72}$$

$$\frac{\partial}{\partial z}I(z,t)+C\frac{\partial}{\partial t}V(z,t)+GV(z,t)=0 \tag{8.73}$$

where:
  I and V are in the 2×1 matrix form
  $R, L, G,$ and C are in the 2×2 matrix form given as

$$R=\begin{pmatrix} R_1 & 0 \\ 0 & R_2 \end{pmatrix}, L=\begin{pmatrix} L_1 & M \\ M & L_2 \end{pmatrix}, G=\begin{pmatrix} G_1 & 0 \\ 0 & G_2 \end{pmatrix}, \text{ and } C=\begin{pmatrix} C_1+C_c & -C_c \\ -C_c & C_1+C_c \end{pmatrix} \tag{8.74}$$

For N number of coupled transmission lines, the I and V are converted into $N \times 1$ matrix and $R, L, G,$ and C into $N \times N$ matrix. All previous recursive relations derived for TSV are applicable for coupled lines by just changing the parameter matrix.

8.3 Performance Analysis of TSVs

As discussed in Chapter 7, TSVs are used for distributing clock signals, power supply, and connecting signals among ICs. In other manner, we can say that TSVs are medium in IC for propagation of signals from one device to other device. Therefore, TSV performs well, if the propagating signal at the far end of TSV reaches at the desired time and desired shape [8–14].

However, after introducing the submicron device technology, a significant change is noticed in TSV performance. Typical time-domain effects include TSV delay, crosstalk, transmission line effects, and noise-on-delay effects highly degraded the performance of TSVs. This chapter presents a detailed time-domain analysis based on the FDTD model developed in Chapter 7 for coupled TSV configuration.

8.3.1 Crosstalk Noise

Aggressive usage of high-speed circuit families, scaling of power supply and threshold voltages, and mixed signal integration combine to make the chips more noise sensitive [15–18]. Crosstalk noise in IC is the result of coupling effects, by which a signal transmitted on one circuit or channel of system creates an undesired effect in another circuit or channel [19–25]. Crosstalk noise can affect the circuit performance in two ways, either by disturbing a quiet line or by changing the response speed of the adjacent switching line.

8.3.1.1 Functional Crosstalk

The current is injected into a quiet line (victim) during the switching of adjacent line (aggressor) is defined as functional crosstalk. An undesirable voltage spike generated at the quiet line (victim) during the switching of adjacent line is defined as functional crosstalk. This type of crosstalk noise can cause device or logic failure due to unexpected voltage peak [26–31].

A simple setup for functional crosstalk analysis is presented here. Figure 8.14 shows the experimental setup for coupled TSV configuration driven by a resistive driver. One of the lines is identified as aggressor and the other one as victim. The driver of aggressor line is excited by a rising ramp source of $V_s = 0.9$ V with 50 ps transient time, in series with a linear resistance $R_s = 75$ Ω. However, the driver of victim line is grounded through a linear resistance $R_v = 50$ Ω.

Coupled TSV wire of 1 mm length is considered for crosstalk analysis. The TSV wire has parameters such as ground capacitance $(C) = 257$ fF, self-inductance $(L) = 2.15$ nH, coupling capacitance $(C_c) = 184$ fF, and mutual-inductance $(M) = 1.68$ nH. TSV is terminated by a load capacitance (C_l) of 10 fF.

The response of victim line at the far end is shown in Figure 8.15. An aggressor line switching results in a voltage spike of 0.234 V at the far end of the victim line. If a device connected at the far end of the victim line has a threshold voltage in this range, a logical failure or false switching of gate may occur.

8.3.1.2 Dynamic Crosstalk

Dynamic crosstalk appears due to the simultaneous switching at adjacent lines either in the same phase or in the opposite phase. This noise impacts the critical issue of timing by changing the response time of the adjacent signal.

Analyzing the dynamic crosstalk noise case is equally important as modeling of functional crosstalk noise, particularly because CMOS logic gates tend to have very good functional noise rejection capabilities, so dynamic case is essential for complete analysis of on-chip performance.

Using simulation setup shown in Figure 8.16, dynamic crosstalk is analyzed, when both the aggressor and victim lines are simultaneously switched

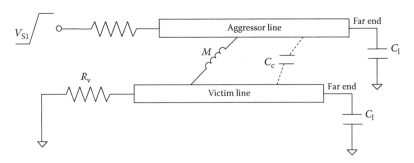

FIGURE 8.14
Coupled *RLGC* TSV configuration for functional crosstalk.

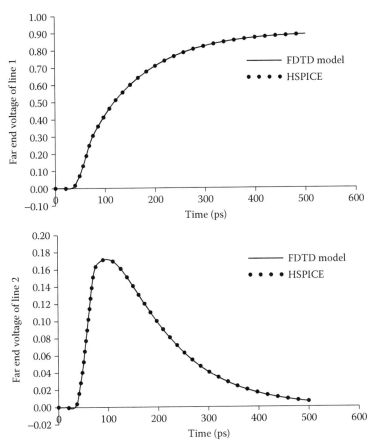

FIGURE 8.15
Comparison of Hewlett simulation program with integrated circuit emphasis (HSPICE) and FDTD-generated waveforms at the far end of the victim line.

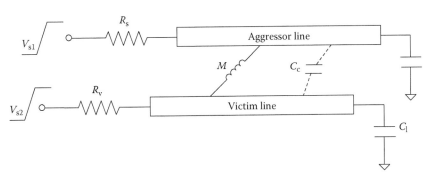

FIGURE 8.16
Coupled line configurations for dynamic crosstalk analysis (in-phase switching).

in the same phase by s rising ramp signal (0-V_s) with 50 ps transition time. The time-domain response is shown in Figure 8.17. The positive peak at the far end of the victim line reaches up to 1.57 V for a 1.2 V input ramp signal.

In the second case of dynamic crosstalk, TSVs lines is excited in out-phase as shown in Figure 8.18. A rising ramp signal is applied at the aggressor line, whereas a falling ramp signal is used for victim line. The time-domain response at the far end of both lines is shown in Figures 8.19 and 8.20, respectively.

8.3.2 Effect of TSV Length Variation

A similar functional crosstalk analysis setup is used. TSV length varies from 1 to 4 mm. In all cases, the line parasitic is calculated, and the FDTD model is implemented for output response. Figure 8.21 shows the variation in noise peak at the far end of the victim line with capacitive load.

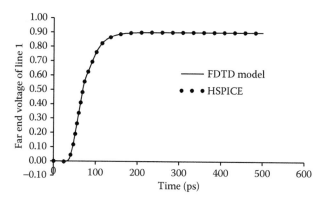

FIGURE 8.17
Comparison of Hewlett simulation program with integrated circuit emphasis (HSPICE) and FDTD-generated waveforms at the far end of the victim line under in-phase switching.

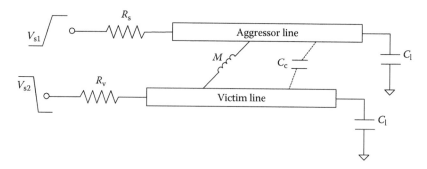

FIGURE 8.18
Coupled line configurations for dynamic crosstalk analysis (out-phase switching).

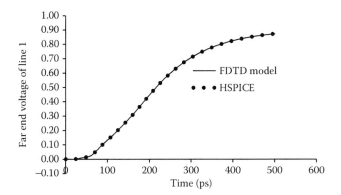

FIGURE 8.19
Comparison of Hewlett simulation program with integrated circuit emphasis (HSPICE) and FDTD-generated waveforms at the far end of the aggressor line under out-phase switching.

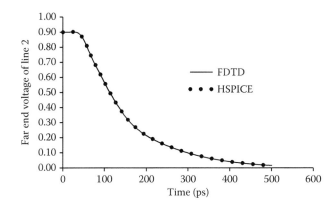

FIGURE 8.20
Comparison of Hewlett simulation program with integrated circuit emphasis (HSPICE) and FDTD-generated waveforms at the far end of the victim line under out-phase switching.

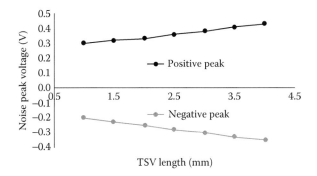

FIGURE 8.21
Noise peak variation with TSV length.

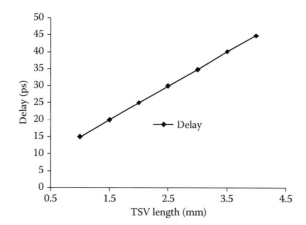

FIGURE 8.22
Delay variation with TSV length.

Figure 8.22 shows the delay variation at the aggressor line end with respect to TSV length. According to Figure 8.22, increment in delay linearly varies with TSV line.

8.4 Summary

This chapter presented a model that accurately computes the performance of TSVs in ICs. Time-domain updating equations are derived using the FDTD technique. An accurate FDTD-based method for TSVs has been modeled using *RLGC* transmission line. The accuracy of FDTD is improved by using appropriate discretization in space and time point. The explicit forms of the boundary conditions are derived from the implicit interface equations without any approximation, and thus, the stability of the proposed method is strictly constrained by the Courant condition. All analyses were carried out on practical problems, and they are supported with simulation results. The measurement results were in good agreement with simulation results, thus leading to the conclusion that all stipulated theories are accurate.

Multiple Choice Questions

1. The stability of the FDTD is limited by
   a. Time step size
   b. Space step size
   c. Both a and b
   d. None of the above

2. In the FDTD model, the boundary conditions are derived in
   a. An explicit manner
   b. An implicit manner
   c. Time domain
   d. None of these

3. The phase velocity of the signal in the transmission line depends on
   a. Transmission line parasitics
   b. Input switching
   c. Coupling parasitics
   d. All of the above

4. In the FDTD model, the voltage and adjacent current solution point is separated by
   a. Δz
   b. $\Delta z/2$
   c. $\Delta z/3$
   d. $\Delta z/4$

5. In dynamic out-phase crosstalk, the delay increases due to
   a. High resistance
   b. High inductance
   c. High capacitance
   d. All of the above

6. In dynamic in-phase crosstalk, the delay decreases due to
   a. Low resistance
   b. Low inductance
   c. Low capacitance
   d. All of the above

7. For closely placed TSVs, the value of mutual inductance is
    a. Equal to line inductance
    b. Greater than line inductance
    c. Less than line inductance
    d. None of the above
8. For closely placed TSVs, the value of coupling capacitance is
    a. Equal to line capacitance
    b. Greater than line capacitance
    c. Less than line capacitance
    d. None of the above
9. Signal integrity of the driver–TSV–load system is primarily affected by
    a. Driver
    b. TSV
    c. Load
    d. All of the above
10. The accuracy of the FDTD model depends on
    a. The space step size
    b. The time step size
    c. The number of iterations
    d. All of the above

Short Questions

1. Define the Courant stability condition.
2. How is the accuracy of the FDTD model dependent on the wavelength?
3. Why is the stability of the FDTD model limited by the Courant condition?
4. What are the differences between implicit and explicit schemes?
5. How can the FDTD model be extended to n-coupled TSVs?
6. Compare the crosstalk-induced propagation delay during the in-phase and out-phase switching of coupled TSVs.
7. How to determine the equal resistance of a nonlinear CMOS inverter?
8. How to determine the maximum time step size in the FDTD model?
9. How to determine the maximum space step size in the FDTD model?
10. Derive the boundary for the FDTD model for a capacitive driver.

Long Questions

1. How to avoid the Courant stability condition in the FDTD model?
2. Analyze the response of three coupled TSVs in all switching cases and find out the worst-case delay.
3. Compare the performance of TSVs using a resistive driver and a CMOS driver.
4. Incorporate the frequency-dependent parameters in the FDTD model.
5. Derive the FDTD model while considering the effect of silicon substrate.

References

1. Yee, K. 1966. Numerical solution of initial boundary value problems involving Maxwell's equations in isotropic media. *IEEE Transactions on Antennas and Propagation* 14(3):302–307.
2. Courant, R., Friedrichs, K., and Lewy, H. 1967. On the partial difference equations of mathematical physics. *IBM Journal* 11(2): 215–234. English translation of the 1928 German original.
3. Li, X., Mao, J., and Swaminathan, M. 2011. Transient analysis of CMOS-gate-driven RLGC interconnects based on FDTD. *IEEE Transactions on Computer-Aided Design of Integrated Circuits and Systems* 30(4):574–583.
4. Paul, R. 1996. Decoupling the multi conductor transmission line equations. *IEEE Transactions on Microwave Theory and Techniques*. 44(8):1429–1440.
5. Bakoglu, H. B. 1990. *Circuits, Interconnections, and Packaging for VLSI*. Indiana, IN: Addison-Wesley.
6. Sakurai, T., and Newton, A. R. 1990. Alpha-power model and its applications to CMOS inverter delay and other formulas. *IEEE Journal of Solid-State Circuits* 25(2):584–594.
7. Kaushik, K., Sarkar, S., Agarwal, R. P., and Joshi, R. C. 2007. Crosstalk analysis of simultaneously switching coupled interconnects driven by unipolar inputs through heterogeneous resistive drivers. In *Proceedings of the IEEE International Conference on Emerging Technologies*, November 12–13, pp. 278–283. Islamabad, Pakistan.
8. Rabaey, J., Chandrakasan, A., and Nikolic, B. 2003. *Digital Integrated Circuits: A Design Perspective*, 2nd edition. Englewood Cliffs, NJ: Prentice-Hall.
9. *International Technology Roadmap for Semiconductors Interconnect*, 2005, 2007 and 2009 editions. http://www.itrs.net.
10. Moll, F., and Roca, M. 2004. *Interconnection Noise in VLSI Circuits*. New York: Springer.
11. Kaushik, B. K., Goel, S., and Rautham, G. 2007. Future VLSI interconnects: Optical fiber or carbon nanotube—A review. *Microelectronics International* 24(2):53–63.
12. Hung, P. S. 2002. *An Interconnect-Driven System-on-Chip Floor Planning Framework*, Ph.D thesis. Stanford University, Stanford, CA.

13. Moll, F., Roca, M., and Rubio, A. 1998. Inductance in VLSI interconnection modeling. *IEEE Proceedings—Circuits Devices System* 145(3):175–179.

14. Dutta, S., Shetti, S. S. M., and Lusky, S. L. 1995. A comprehensive delay model for CMOS inverters. *IEEE Journal of Solid-State Circuits* 30(8):864–871.

15. Bisdounis, L., Nikolaidis, S., and Koufopavlou, O. 1998. Analytical transient response and propagation delay evaluation of the CMOS inverter for short-channel devices. *IEEE Journal of Solid-State Circuits* 33(2):302–306.

16. Qian, J., Pullela, S., and Pillage, L. 1994. Modeling the "effective capacitance" for the RC interconnect of CMOS gates. *IEEE Transactions on Computer-Aided Design* 13(12):1526–1535.

17. Hafed, M., Oulmane, M., and Rumin, N. C. 2001. Delay and current estimation in a CMOS inverter with an RC load. *IEEE Transactions on Computer-Aided Design* 20(1):80–89.

18. Chatzigeorgiou, A., Nikolaidis, S., and Tsoukalas, I. 2001. Modeling CMOS gates driving RC interconnect loads. *IEEE Transactions on Circuits and Systems II: Analog Digital Signal Processing* 48(4):413–418.

19. Kaushik, K., Sarkara, S., and Agarwala, R. P. 2007. Waveform analysis and delay prediction for a CMOS gate driving RLC interconnect load. *Integration, Very Large Scale Integration Journal* 40(4):394–405.

20. Rubinstein, J., Penfield, P., and Horowitz, M. A. 1983. Signal delay in RC tree networks. *IEEE Transactions on Computer-Aided Design* 2(3):202–211.

21. Kahng, B., and Muddu, S. 1997. An analytical delay model for RLC interconnects. *IEEE Transactions on Computer-Aided Design* 16(2):1507–1514.

22. Ismail, Y. I., Friedman, E. G., and Neves, J. L. 2000. Equivalent Elmore delay for RLC trees. *IEEE Transactions on Computer-Aided Design* 19(1):83–97.

23. Sakurai, T. 1993. Closed-form expressions for interconnect delay, coupling, and crosstalk in VLSIs. *IEEE Transactions on Electron Devices* 40(1):118–124.

24. Eo, Y., Eisenstadt, W. R., Jeong, J. Y., and Kwon, O.-K. 2000. A new on-chip interconnect crosstalk model and experimental verification for CMOS VLSI circuit design. *IEEE Transactions on Electron Devices* 47(1):129–140.

25. Ding, L., Blaauw, D., and Mazumder, P. 2003. Accurate crosstalk modeling for early signal integrity analysis. *IEEE Transactions on Computer-Aided Design* 22(5):627–635.

26. Bai, X., Chandra, R., Dey, S., and Srinivas, P. V. 2004. Interconnect couplingaware driver modeling in static noise analysis for nanometer circuits. *IEEE Transactions on Computer-Aided Design* 23(8):1256–1263.

27. Davis, J. A., and Meindl, J. D. 2000. Compact distributed RLC models, part I: Single line transient, time delay, and overshoot expressions. *IEEE Transactions on Electron Devices* 47(11):2068–2077.

28. Davis, J. A., and Meindl, J. D. 2000. Compact distributed RLC models, part II: Coupled line transient expressions and peak crosstalk in multilevel networks. *IEEE Transactions on Electron Devices* 47(11):2078–2087.

29. Agarwal, K., Sylvester, D., and Blaauw, D. 2006. Modeling and analysis of crosstalk noise in coupled RLC interconnects. *IEEE Transactions on Computer-Aided Design* 25(5):892–901.

30. Paul, R. 1994. Incorporation of terminal constraints in the FDTD analysis of transmission lines. *IEEE Transactions on Electromagnetic Compatibility* 36(2):85–91.

31. Ng, K.K., and Lynch, W.T. 1987. The impact of intrinsic series resistance on MOSFET scaling. *IEEE Transactions on Electron Devices* ED-34:503–511.

Answers to Multiple Choice Questions

Chapter 1

1. (c), 2. (b), 3. (d), 4. (d), 5. (d), 6. (d), 7. (a), 8. (c), 9. (d), 10. (d).

Chapter 2

1. (c), 2. (c), 3. (a), 4. (a), 5. (a), 6. (a), 7. (c), 8. (b), 9. (b), 10. (d), 11. (b), 12. (d), 13. (b), 14. (d), 15. (a).

Chapter 3

1. (d), 2. (b), 3. (b), 4. (c), 5. (a), 6. (a), 7. (d), 8. (b), 9. (a), 10. (a).

Chapter 4

1. (b), 2. (d), 3. (a), 4. (c), 5. (b), 6. (c), 7. (b), 8. (d), 9. (a), 10. (c).

Chapter 5

1. (a), 2. (c), 3. (a), 4. (c), 5. (b), 6. (d), 7. (d), 8. (b), 9. (a), 10. (c).

Chapter 6

1. (d), 2. (b), 3. (c), 4. (d), 5. (b), 6. (b), 7. (a), 8. (a), 9. (a), 10. (b).

Chapter 7

1. (b), 2. (a), 3. (c), 4. (d), 5. (c), 6. (b), 7. (a), 8. (d), 9. (c), 10. (a).

Chapter 8

1. (c), 2. (b), 3. (d), 4. (b), 5. (c), 6. (c), 7. (c), 8. (b), 9. (b), 10. (d).

Index

Note: Page numbers followed by f and t refer to figures and tables, respectively.